POPULATION AND FOOD

The world's population is expected to grow by about 2.2 billion people between now and the year 2020. There is no doubt that feeding these extra mouths represents the principal challenge currently facing the global system of agricultural production and trade.

Population and Food examines recent trends in food production and assesses the prospects for feeding humanity up to the twenty-first century. Synthesizing a mass of statistical data and drawing on case material from Africa, Asia, Latin America, Europe, North America and the Middle East, the book suggests that food production in most world regions has kept ahead of population growth. Considering likely future trends in climate, land resources, water availability, trading patterns, farm inputs and technological innovation, the author argues that there should be no insurmountable problems in meeting the world's food demand to the year 2020 and questions the current pessimism voiced about future food prospects.

Tim Dyson is Professor of Population Studies at the London School of Economics.

GLOBAL ENVIRONMENTAL CHANGE SERIES

The *Global Environmental Change Series*, published in association with the ESRC Global Environmental Change Programme, emphasizes the way that human aspirations, choices and everyday behaviour influence changes in the global environment. In the aftermath of UNCED and Agenda 21, this series helps crystallize the contribution of social science thinking to global change and explores the impact of global changes on the development of social sciences.

POPULATION AND FOOD

Global trends and future prospects

Tim Dyson

Global Environmental Change Programme

London and New York

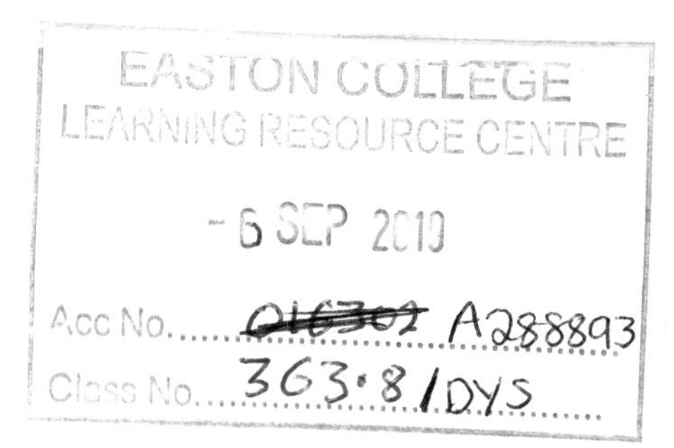

First published 1996
by Routledge
11 New Fetter Lane, London EC4P 4EE

Simultaneously published in the USA and Canada
by Routledge
29 West 35th Street, New York, NY 10001

Typeset in Garamond by
J&L Composition Ltd, Filey, North Yorkshire.
Printed and bound in Great Britain by
Clays Ltd, St. Ives PLC

British Library Cataloguing in Publication Data
A catalogue record for this book is available from the British Library

Library of Congress Cataloguing in Publication Data
Dyson, Tim
Population and food: global trends and future prospects/Tim Dyson.
p. cm.—(Global environmental change series)
Includes bibliographical references and index.
1. Food supply. 2. Population. I. Title. II. Series.
HD9000.5.D97 1996
363.8–dc20 95–38227

ISBN 0–415–11974–X (hbk)
ISBN 0–415–11975–8 (pbk)

For Sue, Tris and Nick,
with much love

CONTENTS

List of figures x
List of tables xii
Preface xiv
Abbreviations xviii

1 PESSIMISTS AND OPTIMISTS
Population and food 1
Some history and theory 3
Some anti-Malthusian views 6
The modern neo-Malthusian case 11
A question of balance 18
Concluding remarks 22

2 POPULATION AND FOOD TODAY
Introduction 24
Data sources and data quality 24
Dividing up the world 28
The definition and extent of hunger 33
Cereal production, consumption, trade and aid 39
Population growth and food supplies 47
An aside on food security 51
Concluding remarks 55

3 POPULATION AND FOOD: RECENT TRENDS
Introduction 58
Trends in world cereal production 58
Where is the 'green revolution'? 64
Food crises and famines 68
Regional cereal production trends 75
Résumé and commentary 83

Changes in harvest variability 87
Trends in food trade and aid 89
Trends in food production, availability and variety 91
Concluding remarks 95

4 EXPLORING THE FUTURE: DEMAND AND SUPPLY
Introduction 100
Future demand for cereals 101
Future supply of cereals 115
Weighing 'demand' and 'supply' 125
Overview of the scenario 132
Concluding remarks 133

5 EXPLORING THE FUTURE: POTENTIALS AND CONSTRAINTS
Introduction 136
Global atmospheric changes 136
Soil and water 143
Farm inputs, crop developments and research 155
Physical and human capital 161
Institutions and markets 163
Food prices 166
Concluding remarks 168

6 EXPLORING THE FUTURE: REGIONAL VIGNETTES
Introduction 170
Sub-Saharan Africa 170
The Middle East 177
South Asia 180
The Far East 184
Latin America 191
Europe and the former Soviet Union 192
North America and Oceania 196
Concluding remarks 199

7 CONCLUSIONS, FORECASTS, CAVEATS, TEMPERED HOPE
Preamble 201
Conclusions 201
Forecasts 203
Caveats 205
Tempered hope 208

Appendix: Country Data and Food Security Index, around 1990 210
Bibliography 212
Index 223

FIGURES

1.1	Simple model to illustrate possible interactions between changes in population density and changes in per capita food production	9
1.2	Population and food: key relationships in the modern neo-Malthusian case	14
1.3	Factors which may increase future food production	21
2.1	Distribution of sample countries by per capita daily calorie supply, around 1990	40, 41
2.2	Population growth rate and per capita calorie supply around 1990, sample countries	49
3.1	World per capita cereal production, 1951–93	59
3.2	World cereal yield and area harvested per capita, 1951–93	61
3.3	Average annual growth rates of world per capita cereal production, area harvested and yield, based on five year averages	63
3.4	Per capita cereal production in selected sample countries, 1951–93	66, 67
3.5	Per cent of developing countries with per capita cereal production above long-run trend, 1951–93	71
3.6	World cereals: prices, stocks and donations, 1960–93	72
3.7	Per capita cereal production by world region, 1951–93	76, 77
3.8	Average annual growth rates of regional per capita cereal production, area harvested and yield, based on five year averages	78, 79
3.9	Growth rates of per capita cereal production and population, 67 sample developing countries	86

4.1 Estimated contribution (per cent) of world population growth to world cereal consumption growth, decades 1950–60 to 1980–90 107
4.2 Assumed trajectories of per capita cereal consumption, 1990–2020 112
4.3 Fertilizer use and cereal yields, sample countries, 1989–92 120
5.1 World average annual temperatures, 1950–93 137
5.2 Growth in world irrigated area, 1961–92 152
6.1 Cereal yields for selected countries in the Far East, 1951–93 189

TABLES

1.1 World population parameters and projections, 1930–2020 12
2.1 Population of main world regions in 1990 and measures of regional representativeness of sample countries 30
2.2 Selected demographic and socio-economic measures, world regions, around 1990 32
2.3 Measures of undernutrition and calorie supply, world regions, around 1990 36
2.4 Average annual measures of cereal production, consumption, trade and aid, world regions, around 1990 43
2.5 Average percentage distribution of total global cereal aid flows in 1989–90 and 1990–91, from major donors to recipient regions 45
2.6 Distribution of sample countries by relative food security status and world regions 54
3.1 Summary measures of average annual world cereal production, selected periods 60
3.2 Regional measures of recent changes in cereal production 81
3.3 Measures of cereal harvest variability, by world region and period 88
3.4 Regional cereal trade and aid expressed as percentage of total regional cereal consumption, 1969–71 to 1990–92 91
3.5 Indices of per capita cereal production, food production and calorie availability, world regions, selected periods 93

4.1 Factors affecting future demand for cereals, world regions, 1990–2020 102
4.2 Measures of cereal consumption growth, world regions, 1970–90 106
4.3 Projected demand/consumption of cereals, world regions, 2020 111
4.4 Factors affecting the future supply of cereals, world regions, around 1990 116
4.5 Percentage contributions of area and yield to total cereal production growth, and growth rates of harvested cereal area, world regions, 1970–90 121
4.6 Average cereal yields, levels, trends and projections, world regions, 1951–2020 123
4.7 A plausible 2020 scenario: reconciling projected regional cereal demand and supply 126
5.1 Sample countries with declining cereal yields between 1981–86 and 1987–92 147
6.1 Average measures of cereal production within sub-regions of Europe/FSU, 1990–92 194

PREFACE

> Food not only constitutes the principal part of the riches of the world, but it is the abundance of food which gives the principal part of their value to many other sorts of riches.
>
> (Adam Smith, 1776, pp. 218–19)

> In order to know anything it is necessary to know everything, but in order to talk about anything it is necessary to neglect a great deal.
>
> (Joan Robinson, 1941, p. 8)

I must be a little crazy to have written this book. The subject is so large. But in the early 1990s I read a claim that the world's population was growing faster than the production of grain and, more striking still, that population growth was now outpacing grain production in each major geographical region. These statements have been followed by other disturbing assertions – for example, that growth in the average world grain yield is now dramatically slowing down.[1] And in the mid-1990s there is certainly no shortage of writers who argue that humanity will soon face really colossal food problems. This dire perspective is sometimes termed 'neo-Malthusian', because in its emphasis on hunger, starvation and famine it echoes themes popularized by Thomas Robert Malthus' polemical essay of 1798.[2]

[1] The writings of Lester R. Brown are particularly prominent on these issues. See, for example, Brown (1991, pp. 11–15; 1994a, pp. 177–87) and Brown and Kane (1995, especially chs 2 and 10).

[2] The full title of this book, which achieved immediate notoriety, is *An Essay on the Principle of Population as it Affects the Future Improvement of Society with Remarks on the Speculations of Mr. Godwin, M. Condorcet, and Other Writers.* See Malthus (1798).

Hence the research which has led to this book. The subject of world food prospects is so huge that its assessment cannot depend upon any single methodology or discipline. I certainly do not know everything. I have certainly had to neglect a great deal. But as the work has progressed so I have benefited from a growing belief that a non-agricultural 'outsider' (like me) may stand as good a chance of reaching a sober assessment as do many insiders – who perhaps are too intimately, or have been too long, involved.

Anyhow, for the present book I have read widely. I have trawled through an ocean of statistics. I have consulted many individual experts (although in all cases the usual disclaimer applies). And, not least, I have talked to farmers in a host of different locations – like the East African highlands, the Cauvery and Red River deltas, the Honde and Jordan valleys, and the flat, productive wheat fields of eastern England.

Of course, the upshot of all this is an amalgam. And ultimately the issue is a matter of judgement. Certainly, some of my detailed speculations herein will not prove correct. But, this said, I am fairly confident of the 'big picture'. The principal conclusion of this book is that the alarming neo-Malthusian prognosis for the world during the next two or three decades is almost certainly wrong.

For the world and most of its regions the overall balance between population and food has not deteriorated in the recent past. In fact, there have generally been improvements in average levels of per capita food consumption. There is no reason to believe that growth in the global cereal grain yield is seriously faltering. And there is fair reason to expect that in the year 2020 (the principal time horizon for the present book) world agriculture will be feeding the then larger global population no worse – and probably a little better – than it manages to do in the mid-1990s.

However, the problem with penning the previous paragraph is that it immediately opens me to the rather sweeping charges of 'optimism' and 'complacency'. I know nobody who would admit to the latter. But for many people who write about world food prospects complacency is seen as a widespread and heinous sin.

So let me state quite clearly that there are indeed significant difficulties and dangers ahead. Population growth will pose the main challenge for world food production in the next few decades; slower demographic growth would make the situation easier to handle. The somewhat better feeding of humanity in the year 2020 can really be accomplished only by a considerable increase in the volume of inter-regional food transfers through trade. In parts of sub-Saharan Africa there are neo-Malthusian resonances. There is strong evidence of an increasing frequency of drought in some world regions. The risk of another global food crisis, such as occurred in 1972–74 (actually relatively modest in scale, but still very painful and avoidable) cannot be entirely ruled out.

So the basic theme of promise in this book is nevertheless restrained. The picture is mixed. You, the reader, will soon appreciate these and other important caveats which I wish to make. Few areas of human activity represent a more worthwhile sector for investment now than does food and agricultural production. Gains in food output can underpin so much else besides. Hopefully it will not need a crisis to remind us of this.

Most of the work for this book has been conducted during a two year Research Fellowship held under the Global Environmental Change (GEC) Programme of the UK Economic and Social Research Council (ESRC Award L320-27-3024). I am immensely grateful for the intellectual freedom and resources to travel, see and consult which this fellowship has provided. In addition, an earlier period of sabbatical leave from the London School of Economics allowed me to lay the groundwork for this study.

Several institutions and research groups have been particularly helpful in assisting this outsider to establish contact with insiders in the world of food and agriculture. The early-morning BBC radio programme *Farming Today* has been a tremendous resource. I would also like to thank members of the World Hunger Program at Brown University, the Food Security Unit at the Institute of Development Studies in Sussex, the International Centre for Research in Agroforestry in Nairobi and the International Rice Research Institute in Manila.

In addition, in these and many other institutions many people have assisted me in various ways. For help and advice I am grateful to Benjamin Amadalo, Tenkir Bonger, Diana Callear, Robert Cassen, Ken Cassman, Tang Zhu Chang, Fang Bing Chu, the late Nigel Crook, Monica Das Gupta, Mahendra Dev, Jean Drèze, Jiao Bi Fang, Simon Gregson, David Grigg, Mohamad Hamdan, Hasan Bin Ismail, Nurul Islam, Jimmy Kiio, the late Gopu Kumar, Amitabh Kundu, Richard Leete, Mohamad Ma'ani, Jay Maclean, Simon Maxwell, George Mburathi, Ellen Messer, Behrooz Morvaridi, Mina Moshkeri, Paul Muchena, Godfrey Mudimu, Adrian Mukhebi, Mike Murphy, Sandra Postel, Michael Redclift, Robin

Reid, Rabbi Royan, Lawrence Rubey, Xia Shun-Kang, Jayne Stack, Rob Swinkels, Sant Virmani, Harry Walters, Peng Xizhe, Li Ming Yun and Christopher Zindi, among others. All my immediate colleagues at LSE have provided help and encouragement. And I have been particularly fortunate in having the really splendid secretarial and administrative assistance of Marilyn Hynes.

The book is dedicated to my family who, as usual, have had to put up with a lot. The royalties will go to OXFAM.

Lastly, a word on organization and presentation. The first two chapters essentially set the scene. Perhaps the core of the book lies in chapters 3 and 4, which examine recent trends and present some simple calculations about the future. Chapter 5 addresses the context and basis of future world food production. Chapter 6 examines regional prospects. Finally, chapter 7 briefly summarizes the main conclusions and attempts a balanced discussion of the way ahead. The text is fairly informal. The subject of the book is important. So I have endeavoured to make the contents accessible to as many readers as possible.

Tim Dyson

ABBREVIATIONS

ha	hectare (2.471 acres)
kg	kilogram
mt	metric ton (1000 kg) – all tons are metric tons (mt)
$US	United States dollar
BMR	Basal Metabolic Rate
CAP	Common Agricultural Policy (of the European Union)
CFC	Chlorofluorocarbon Gas
CGIAR	Consultative Group on International Agricultural Research
CIMMYT	Centro Internacional de Mejoramiento de Maiz y Trigo
EC	European Community – now the European Union (EU)
ENSO	El Niño–Southern Oscillation
EU	European Union – formerly the European Community (EC)
FAD	Food Availability Decline
FAO	Food and Agriculture Organization (of the United Nations)
FSI	Food Security Index
FSU	Former Soviet Union
GATT	General Agreement on Tariffs and Trade
GDP	Gross Domestic Product
GHG	Greenhouse Gas
GIEWS	Global Information and Early Warning System
GLASOD	Global Assessment of Soil Degradation
GNP	Gross National Product
HIV/AIDS	Human Immunodeficiency Virus/Acquired Immune Deficiency Syndrome

HYV	High Yielding Variety
ICRAF	International Centre for Research in Agroforestry
ICRISAT	International Crops Research Institute for the Semi-Arid Tropics
IFPRI	International Food Policy Research Institute
IMF	International Monetary Fund
IPCC	Intergovernmental Panel on Climate Change
IPM	Integrated Pest Management
IRRI	International Rice Research Institute
NAFTA	North American Free Trade Agreement
NGO	Non-Governmental Organization
UK	United Kingdom
UN	United Nations
UNEP	United Nations Environment Programme
US	United States
USDA	United States Department of Agriculture
UV-B	Ultraviolet-B
WHO	World Health Organization
WTO	World Trade Organization

1

PESSIMISTS AND OPTIMISTS

POPULATION AND FOOD

This book is about the relationship between population and food – in the past, the present and the future. Its single most important aim is to assess the prospects for this relationship during the next 20 to 30 years, as the world's population steadily grows by nearly 900 million people each decade.

No small task! Inevitably, we must simplify and abstract. And no one can be certain of the future. Yet there are several considerations which help to make the task more manageable and which probably increase the chances that our assessment will prove to be fairly realistic, at least in general terms.

First, interest in the relationship between population and food – and in particular, anxiety about the *balance* of this relationship – is far from new. Indeed, there is a great mass of relevant theory, analysis, experience and debate on which we can draw.

Second, any assessment of future prospects for the relationship will undoubtedly profit from a close examination of recent trends and the current situation. The future may be uncertain. But we can be certain that it will be heavily conditioned by both present circumstances and trends during the recent past.

Finally, our assessment will benefit if we recognize from the outset that the relationship between population and food can be viewed as both very *simple* and very *complex*.

In its simple form the relationship can be treated as approximately an identity between two quantities. For example, between 1800 and 1930 the world's population roughly doubled in size from about 1 billion to 2 billion people. It is therefore reasonable to assume that the total amount of food produced and consumed in the world also probably doubled – more or less. Likewise, if a given quantity of food is available and the size of the population is known, then it is an easy matter to calculate the average amount of food that is available per head. This simple approach to the relationship – essentially roughly equating 'people' with 'food' – is probably as old as humankind. It is nicely illustrated by the Chinese characters which, taken together, correspond to the English word for 'population':

人口

A person is represented on the left. An open mouth, requiring food, appears on the right. Because this perspective focuses on just two quantities, it tends to promote concern about the changing balance (or equilibrium) between population and food. Moreover, it is this simple approach which usually occupies the public mind.

However, the relationship can also be viewed with the whole of human society essentially mediating between the two entities of population and food. Obviously, this is an infinitely more complex approach. It involves, for example, explicit recognition of the fact that social and economic inequalities often determine differences in people's access to food. In virtually no human population is food distributed in exactly equal shares; this is true whether we consider individual families or the world as a whole. Moreover, consider the multitude of ways in which the world's people obtain their food today. At one extreme, many poor households in Africa, Asia and Latin America still depend directly upon their own labour to cultivate, fish, hunt or gather the bulk of the food they eat. At the other extreme, most people living in contemporary western societies have virtually no contact with the original sources of their food; instead, a trip to the local supermarket is sufficient to reveal a huge diversity of foodstuffs from many locations, transported to them and mediated by a vast and complicated system of international trade. Between these two extremes an infinite variety of combinations of arrangements link the world's people to their daily bread.

Both the simple and complex approaches are needed in this book. It will often be necessary to divide quantities of 'food' by numbers of 'people' in order to obtain per capita measures of food availability – for countries, for regions, and indeed for the world as a whole. But we must not become distracted by this simplicity. In particular, we mustn't forget that people – and the societies they comprise – are also the principal producers of the food.

With this as background, the rest of this chapter focuses on the two main theoretical – and indeed psychological – perspectives which have dominated discussions about population and food. Rather simplistically, these are sometimes labelled the 'pessimistic' and 'optimistic' schools of thought. We will give somewhat greater attention to the pessimistic perspective which is often associated with the name of Thomas Robert Malthus. We do this because the pessimistic school tends to make the running in much of the public debate. Indeed, the present study is partly an evaluation of current 'neo-Malthusian' claims that the world is facing an unprecedented crisis of hunger and famine mainly because of population growth. In fact the modern clash between pessimists and optimists forms part of a very old debate. Therefore it is appropriate to start with a soupçon of history.

SOME HISTORY AND THEORY

Malthus was not the first person to suggest that population growth might outstrip the capacity to produce food. Nor was he the first to suggest that this might limit population size by raising death rates.[1] But he was certainly the most influential and impressive early student of the subject. And modern writers of the pessimistic school still frequently acknowledge their debt to him.

It is the clarity, apparent rigour and stimulating expression of Malthus' *An Essay on the Principle of Population*, published anonymously in 1798 (hereafter termed simply the *First Essay*), which caused it to receive such great attention. The basic argument can be sketched with the help of a few famous quotations:

> Population, when unchecked, increases in a geometrical [i.e. compound] ratio. Subsistence [i.e. food production] increases only in an arithmetical ratio . . . By that law of our nature which makes food necessary to the life of man, the effects of

[1] Nor did Malthus claim any great originality: 'The most important argument that I shall adduce is certainly not new.' See Malthus (1798, p. 8). Notable among his eighteenth-century predecessors in exploring such issues were the Dane, Otto Diederich Lütken (see Saether 1993) and the Scot, James Steuart (see Blaxter 1986).

> these two unequal powers [of population and food] must be kept equal.
>
> This implies a strong and constantly operating check on population from the difficulty of subsistence. This difficulty [of providing sufficient food] must fall somewhere and must necessarily be severely felt by a large portion of mankind.[2]

Malthus then proceeded to argue that the balance between population and food is chiefly maintained by upward pressure on the death rate – and that this 'absolutely necessary consequence' was primarily brought about by hunger, starvation and disease. But, should these three factors still be insufficient to restore equilibrium, then:

> gigantic inevitable famine stalks in the rear, and with one mighty blow levels the population with the food of the world.[3]

This is powerful stuff! But then Malthus was deliberately being provocative. The *First Essay* was written specifically to counter some of the extremely optimistic views about the future evolution of human society which were then being expressed by writers influenced by the high aspirations of the French Revolution.[4] Note too that Malthus' focus on the two seemingly simple categories of population and food made it rather easy for him to raise the spectre of a population outgrowing its food supply – provided that the available supply of agricultural land was fixed.

In fact, Malthus' subsequent life-long study of the relationship between population and food – commencing with the publication of his so-called *Second Essay* in 1803 – led him to significantly revise the simple argument of the *First Essay*. He gave increasing importance to changes in the birth rate as an alternative way of maintaining the balance between 'population' and 'food'. Essentially he argued that when food was scarce and times were hard, prices rose and men and women were forced to delay their marriages. In turn, this reduced the birth rate and helped to restore the equilibrium

[2] See Malthus (1798, p. 14). Expressions appearing in square brackets in quotations have been added to assist interpretation.

[3] See Malthus (1798, p. 140). Here the word 'levels' is being used in the sense of 'equalizes'.

[4] The names of two of these optimistic writers – the French philosopher and politician, Jean-Antoine-Nicolas Caritat, Marquis de Condorcet (1743–94) and the English philosopher and novelist, William Godwin (1756–1836) – are actually mentioned in the full title of the *First Essay*. See Malthus (1798).

between population and resources (including food supplies).[5] Also, he realized that trade-offs could occur. For example, a decline in the birth rate would reduce the risk of a rise in the death rate.

Malthus' later writings reveal a very careful scholar, who placed great weight upon the collection and study of empirical data. He used results from the early censuses of the United States to show that in conditions of abundant supplies of land a population could double in size in about 25 years.[6] But he was aware that, because of the operation of various checks, most human populations did *not* actually grow so fast, nor did they usually grow at a geometric (i.e. compound) rate.

Malthus' description of the rate of increase of food production as arithmetic (i.e. linear) was a reasonable generalization given the evidence that was available to him.[7] For present purposes it is important to note that he was discussing the increase of food on the implicit assumption that the supply of cropland was fixed. That is, in modern parlance he was describing the form of growth of agricultural *yields* – e.g. food output per hectare. We shall see in chapters 3 and 4 that aggregate yields have indeed increased in a fairly linear form during recent decades. So, especially the later writings of Malthus on the adaptabilities, flexibilities and general complexities of the relationship between population and food contain many valuable insights which are still valid today.

Since Malthus' time there have been periodic 'waves' of people who have doubted the capacity of the world to produce enough food for its growing population. In a famous speech titled 'The Wheat Problem', Sir William Crookes warned the 1898 meeting of the British Association for the Advancement of Science that unless crop yields were improved the world's growing population would face widespread starvation by the 1930s. The simple argument of the *First Essay* also made a great impression upon John Maynard Keynes, who in *The Economic Consequences of the Peace* took an

[5] Wrigley and Schofield (1981) have demonstrated that such a system did operate in England during the sixteenth, seventeenth and eighteenth centuries. Food is obviously a particularly important resource, and in theoretical discussions it is sometimes treated as an admittedly imperfect proxy for resources in general.

[6] See Malthus (1830/1970, pp. 226–33). The relationship between population growth and food supplies was also a matter of considerable interest to early leaders of the United States. For example, Thomas Jefferson mentioned Malthus' *Second Essay* when writing to the eminent French economist, Jean Baptiste Say, in 1804. See Jefferson (1804/1993).

[7] On this see Flew (1970, p. 19 and p. 34).

essentially neo-Malthusian, pessimistic view of Europe's future following the First World War. The period immediately after the Second World War was another characterized by similar anxiety.[8]

However, before we set out the arguments of the current crop of neo-Malthusian writers we must first introduce those of the opposing, more optimistic, so-called 'anti-Malthusian' school.

SOME ANTI-MALTHUSIAN VIEWS

Instead of seeing a large and growing population as a problem for systems of food production, the opposing perspective sees these characteristics as both a sign and, indeed, a *cause* of prosperity. Again, arguments and opinions which are advanced today chime in well with views expressed long ago. Thus, for the modern American economist Julian L. Simon:

> The ultimate resource is people – skilled, spirited, and hopeful people who will exert their wills and imaginations for their own benefit, and so, inevitably, for the benefit of us all.[9]

And for the Danish economist Friderich C. Lütken, writing in 1756:

> It is my opinion . . . that there can never be too many people in a country . . . people and the multitude of people are the greatest and most splendid wealth by which . . . all other kinds of wealth can be achieved.[10]

An important strand in this optimistic perspective – which can be traced back to the free trade arguments of classical economists like Adam Smith – is the belief that the flexibility and adaptability provided by efficient markets can help solve any problems posed by population growth. Indeed, since population growth increases the volume of demand for food, if scarcity occurs then prices will rise and eventually farmers will be stimulated into producing more food. This leads to the view that, over the long run, population

[8] Keynes wrote: 'Malthus disclosed a Devil. For half a century . . . he was chained up and out of sight. Now perhaps we have loosed him again.' See Keynes (1919, p. 8). For a neo-Malthusian perspective following the Second World War see Huxley (1949). And for a general review of successive neo-Malthusian waves see Kellman (1987, ch. 1).

[9] See Simon (1981, p. 348). Simon – perhaps the leading contemporary exponent of this view – is very open in acknowledging his intellectual predecessors.

[10] For the full quotation see Saether (1993, p. 513).

growth actually causes food to be both more plentiful and cheaper.[11]

But while markets may be important facilitators, there remains the question of the ways in which farm productivity can be raised so as to meet the increased demand for food. Here, two further interrelated strands of the more optimistic anti-Malthusian perspective can be identified.

The first is *science* – or, more specifically, the fruits of agricultural research and their dissemination through increased knowledge and education. It has only really been in the twentieth century that scientific research has led to increases in agricultural productivity. Major advances here include the development of mass production techniques for nitrogen fertilizers (through the Haber–Bosch synthesis of ammonia) around 1913, and the development of so-called 'green revolution' high yielding varieties (HYVs) of wheat and rice during the 1960s. But, writing with foresight in 1844, Friedrich Engels specifically identified science as a key factor which – by supposedly also having the capacity to grow in geometrical progression – could help unhook population growth from the Malthusian constraint of a limited supply of land.[12]

The second main way by which farm productivity may be increased is through the stimulation of agricultural *innovation* induced by population growth itself. In his later writings even Malthus seems to have implicitly accepted the idea that population pressure can lead to the improvement of agricultural techniques and productivity – although he doubted whether such improvements could go very far.[13] But nowadays this idea is most closely associated with the arguments advanced by the economist Ester Boserup in her classic text *The Conditions of Agricultural Growth* published in 1965.

Drawing upon African and Asian data Boserup noted the posi-

[11] See Simon (1981, pp. 62–75; 1992, ch. 1). Of course, unregulated markets have well-known limitations as guarantors of steady supplies of food – limitations which are touched on in chapters 3 and 5 of this book.

[12] Engels wrote that 'science increases at least as fast as population. The latter increases in proportion to the size of the previous generation. Science advances in proportion to the knowledge bequeathed to it by the previous generation, and thus under the most ordinary conditions it also grows in geometrical progression – and what is impossible for science?' For this quotation see Flew (1970, p. 35). As we note below, in some respects scientific contributions to agriculture can indeed be regarded as induced by population growth.

[13] See, for example, Malthus (1830/1970, p. 239).

tive correlation which existed between population density and the intensity of land use. Societies living at low population densities tended to practise 'extensive' agriculture; farmers would cultivate a plot of land with a few simple tools and then move on to work another patch, letting the original plot lie fallow for perhaps 20–25 years during which time its productivity would recover. Then the rejuvenated plot would be cultivated again. However, societies living at comparatively high population densities were much more likely to cultivate the same plot of land each year (i.e. practise annual cropping). These communities tended to have more sophisticated agricultural systems and farm tools. Moreover, levels of food production per person were sometimes higher under such more 'intensive' farming regimes, particularly if greater inputs of agricultural labour per person were required (i.e. in terms of time spent in cultivation).

From this essentially cross-sectional observation Boserup argued that through the challenges to society which it posed, population growth often actually worked as the major dynamic engine of agricultural change – stimulating, in particular, the adoption of improvements in land use and technology. Moreover, this basically anti-Malthusian perspective can be further developed to argue that – other things being equal – the larger a population is, the larger will be the number of farmers, and therefore the greater will be the chance that someone will discover a new and more productive way of cultivating the available supply of land.[14]

There is no doubt that historically such Boserupian processes have been very significant in societies all around the world. Indeed, demographic increase may well have been the single most important cause of agricultural intensification. Furthermore, potentially there are many ways in which food production can be increased to keep up with population growth.[15]

A useful illustration of how such processes might work can be

[14] For the development of these arguments see especially Simon (1992). In fairness to Boserup it must be noted that she realized that her basic argument might not apply in rural locations where population density and growth rates are high. See Boserup (1965, p.118).

[15] Some ways are very direct – e.g. population growth can obviously provide greater supplies of both human labour and night soil (i.e. human excrement for use as fertilizer). But most ways are indirect – e.g. through the substitution of more productive for less productive food crops, the introduction of new crops, and the development of better farming methods. For a panoply of such possibilities see Bilsborrow (1987).

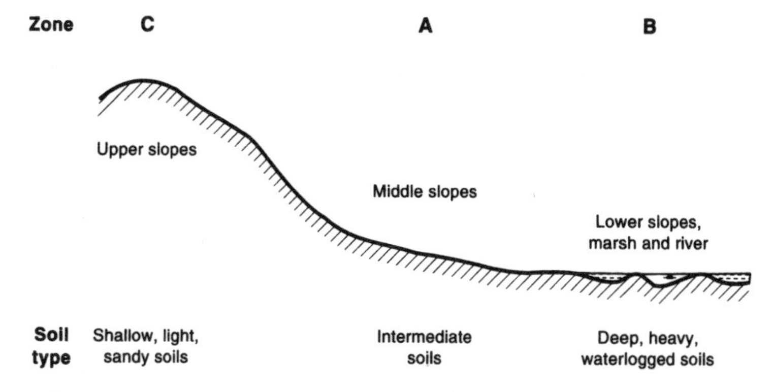

Figure 1.1 Simple model to illustrate possible interactions between changes in population density and changes in per capita food production
Source: Derived from Pingali and Binswanger (1991, p. 57)

derived from a simple model proposed by Pingali and Binswanger. Figure 1.1 shows a hypothetical valley in which the nature of the soil varies according to the location. Quite plausibly, the soils on the upper slopes are light and sandy, whereas those on the valley floor are deep and heavy. A simple farming community moving into the valley for the first time might well decide to settle on the middle slopes (zone A) where the soils are neither too shallow, nor too deep, and can readily be worked by hand. But in time the population of this community might increase, so that it is no longer possible to grow sufficient food in zone A under the prevailing agricultural conditions. At this stage the community might start to work the waterlogged soils of the valley floor (zone B). The larger population now provides enough labour to drain and irrigate the deep, heavy soils of these bottomlands. And once they can be worked and irrigated, such soils often prove to be extremely fertile and well suited, for example, to the intensive cultivation of rice. Food output per person may well increase.

However Figure 1.1 can also illustrate the possible failure of agricultural innovation to cope successfully with demographic growth. The transition from zone A to zone B may in practice be both long and hard. Eventually the population in zone B may increase so that, again, it is no longer possible to grow sufficient food under the prevailing agricultural conditions. In the absence of further innovation, food production per person may start to decline. This can be a real threat. For example, such a scenario is strongly reminiscent of what occurred in densely populated areas of Indonesia during the nineteenth and early twentieth centuries. Clifford Geertz describes a process of 'agricultural involution' in which Javanese systems of wet rice cultivation had to absorb ever-increasing numbers of people. Plots were worked harder and harder, but with little improvement in the methods of cultivation. Food production per person did not increase. 'The process resembles nothing else so much as treading water.'[16]

So our now densely settled communities in zones A and B may confront the possibility of declining levels of per capita food output. The remaining option afforded by Figure 1.1 is migration to the upper slopes. But by their very nature the shallow, light, sandy soils of zone C tend to be low in crop nutrients. Therefore the yields of foodcrops grown on such soils will probably be low. Moreover, upland soils are often fragile and highly erodible. It is an unfortunate fact that the world today is full of cases – e.g. in hill areas of South-east Asia, East Africa and Central America – where precisely this kind of migration to the 'upper slopes' in search of agricultural land has contributed to environmental deterioration, especially soil erosion.

So to sum up, there is no doubt that Boserupian adaptive processes have been very important over the long sweep of human history. But it is likely that they have often operated in a painfully slow and fitful way. Moreover, while increased population density may eventually lead to greater intensification, it is not obvious that this will necessarily be accompanied by increased levels of food production per head. Beyond these considerations, a major limitation of Boserupian processes in relation to conditions in the world today is the sheer speed of contemporary population growth. For example, some populations in sub-Saharan Africa are currently increasing at rates which mean that they will double in size in only about 20 years. There is widespread agreement among agricultural economists that the innovations which poor farmers in much of the developing world are likely to be able to generate by themselves will be quite incapable of keeping up with such very high rates of population growth.[17] Instead, it will be essential to

[16] See Geertz (1963, p. 78).

[17] Thus Pingali and Binswanger (1991, p. 52) write: 'Farmer-generated technical change . . . appears, however, to be incapable of supporting rapidly rising agricultural populations.' See also Boserup (1965, p. 118) and Hayami and Ruttan (1987).

incorporate the fruits of what we previously called 'science' into the overall processes of adaptive response.

But we are getting ahead of ourselves! We have already had cause to mention possible environmental constraints to increases in food production. Since such constraints are a key part of the modern neo-Malthusian case we will turn to consider it now.

THE MODERN NEO-MALTHUSIAN CASE

The two centuries since Malthus penned his *First Essay* have actually provided little evidence to support its classic simple argument. The world's population is certainly both larger and better fed now than it was two hundred years ago. In spite of repeated expressions of anxiety, world food production has generally kept ahead of population growth. In large measure this has been due to the great expansion of the area of land in cultivation (e.g. through the opening-up of North America and Australia), plus the continual development of agricultural practices by farmers and, more recently, the application of modern scientific and industrial approaches to the tasks of raising crop yields and transporting and storing food.[18]

Modern neo-Malthusian writers do not deny that these long-term improvements in per capita food production have taken place. But they argue that in the 1980s and 1990s the world has at last reached a critical turning point. They maintain that in certain crucial respects the situation which humanity now faces is different from that it has confronted in the past. In particular, the neo-Malthusian case highlights the extraordinary scale of contemporary world population growth and the related – and increasingly serious – environmental problems to which this growth contributes. The argument acknowledges that issues of distribution – especially the existence of widespread poverty in much of the developing world, coexistent with overconsumption and waste of resources by rich nations – are very important. But it retains at its core grave anxiety about whether food production can possibly keep up with population growth. Prominent writers such as Lester R. Brown, Paul R. Ehrlich and Norman Myers are forceful proponents of this perspective. Therefore with the help of Figure 1.2 we draw from their writings to summarize key parts of the argument.[19]

[18] For an excellent general review see Grigg (1993). See also Boserup (1990).

[19] See, for example, Brown (1988, 1991, 1993, 1994a), Brown and Kane (1995), Ehrlich and Ehrlich (1990), Ehrlich *et al.* (1993) and Myers (1991).

Table 1.1 World population parameters and projections, 1930–2020

Year	*Population (million)*	*Increment (million)*	*Average annual* Birth rate (per 1000 population)	*Average annual* Death rate (per 1000 population)	*Average annual* Population growth rate (% per year)
1930	2000				
		516	38.6	27.6	1.1
1950	2516				
		503	36.6	18.5	1.8
1960	3019				
		678	34.6	14.4	2.0
1970	3697				
		747	29.6	11.4	1.8
1980	4444				
		841	27.3	10.0	1.7
1990	5285				
		873	24.4	9.1	1.5
2000	6158				
		874	21.6	8.4	1.3
2010	7032				
		856	19.4	7.9	1.1
2020	7888				

Note: The figures shown above for the years 2000, 2010 and 2020 are those of the United Nations 'medium variant' population projections. The 'high' and 'low' UN projections (which assume global birth rates of respectively 22.0 and 16.3 for the period 2010–20) give projected world population totals of 8392 and 7372 million in the year 2020. These other variants are briefly discussed in chapter 4

Principal source: United Nations (1994)

The basic demographic facts underlying the modern neo-Malthusian case are not in dispute. If it took about 130 years for the human population to grow from 1 billion to 2 billion, Table 1.1 shows that the next billion people were added in just 30 years, i.e. by about 1960. The rate of demographic growth has been declining since the 1960s because the world's birth rate has been falling faster than the death rate. But even so, during the 1980s alone the world's population grew by an estimated 841 million people. Moreover, most population projections broadly agree that in the next three decades up to the year 2020 – the principal time-horizon adopted in the present study – the absolute increment will be about 870 million people in each decade (see Table 1.1). The phenomenon of demographic momentum makes most of this growth virtually certain.[20] If it was not until around the year 1800 that the world's population first reached 1 billion, we are currently adding an extra billion people about every 12 years! This unprecedented demographic increase is overwhelmingly located in the developing world where birth rates are highest and population age-structures are youngest. In fact, over 95 per cent of the 2.603 billion increment which the United Nations projects for the period 1990–2020 is expected to occur in the developing countries. This increase in population will be the main cause of a tremendous expansion in the demand for food, and therefore of greatly increased exploitation of land for agricultural production (see Figure 1.2).

At this point environmental considerations and judgements enter the argument and things suddenly become much more complicated. Part of the modern neo-Malthusian case is that food production all over the world is increasingly coming up against various major constraints, of which perhaps the two most important relate to land and water (see Figure 1.2). The total amount of cropland is not changing very much and therefore the quantity per person is shrinking because of population growth. The greater and greater demands which are being made upon the land which is in cultivation are leading to widespread land degradation – in such forms as soil erosion, compaction and desertification. It is claimed that throughout much of the developing world mounting population pressures are directly threatening the inherent productivity of the soil.[21] Fast-growing cities are gobbling-up prime agricultural land, and the dwindling supplies of new land which might be brought into cultivation are generally of much poorer quality.

Turning to water, the problems seem to be just as great. More than one-third of current world agricultural production is said to come from the approximately 16 per cent of agricultural land that is irrigated. But 'Worldwide the prospects for major gains in irrigated area are not good.'[22] So the amount of irrigated cropland

[20] The world's population still has a relatively young age-structure. This means that there is considerable 'built-in' potential for future growth (i.e. demographic momentum) much of which will occur irrespective of how fast the global birth rate declines. For present purposes the World Bank's population projections are sufficiently similar to those of the United Nations (UN) as to not merit special attention. See McNicoll (1992).

[21] See, for example, Brown (1988, pp. 17–22) and Brown and Kane (1995, pp. 148–50).

[22] The quote is from Brown (1988, p. 28). See also Postel (1992, p.49).

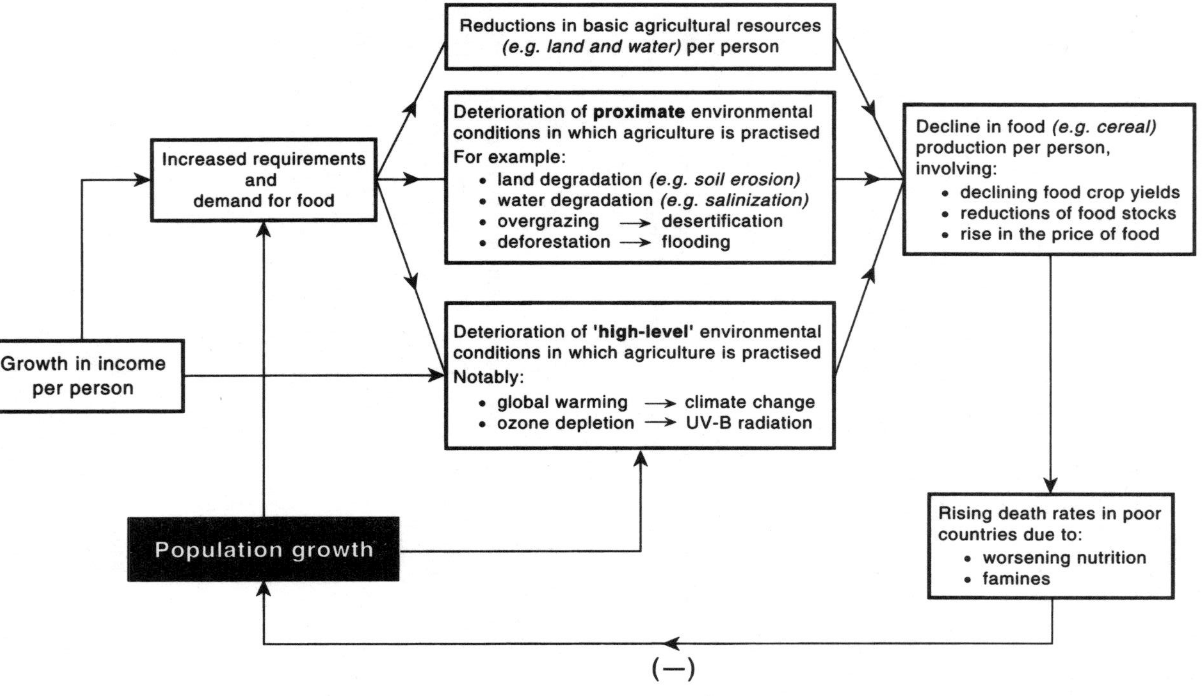

Figure 1.2 Population and food: key relationships in the modern neo-Malthusian case

per person is also declining because of population growth. Again, there are many other difficulties relating to irrigation water than just that of falling per capita supplies. The list includes rising levels of pollution, increased salinization of irrigated croplands, falling water tables and increased competition for water – for example between farmers and city-dwellers, and between nation states. Furthermore, as is often the case apropos environmental problems, these processes frequently combine and interact in synergistic ways. Fast-growing cities threaten supplies of both agricultural land and water. Deforestation of upland areas – often fuelled in part by the migration of poor farmers in search of cultivable land – may cause increased flooding and agricultural destruction downstream (see Figure 1.2).

To the foregoing comparatively down-to-earth challenges facing the world's food production system, the contemporary neo-Malthusian argument proceeds to add several 'high-level' global environmental threats. The most important of these dangers is generally agreed to be global warming, which may threaten food production by inundating low-lying coastal lands (e.g. in the Netherlands, Egypt or Bangladesh) as the level of the sea rises. However, an even greater challenge arising from global warming may be the uncertainty and costs of climate change. 'Farmers, who have always had to deal with the vagaries of weather, now also face the unsettling prospect of climate change.' The agricultural systems of the world 'will be faced with the stresses of continual adaptation to rapidly changing conditions.'[23] Another danger of this type is posed by the depletion of the stratospheric ozone shield surrounding the earth. This may damage important crops like rice and wheat by exposing them to increased levels of harmful ultraviolet-B (UV-B) radiation (see Figure 1.2).

Although threats to food production such as global warming and ozone depletion have largely been caused by the release of greenhouse gases and ozone-destroying chemicals by people living in the more developed regions of the world, this certainly does not mean that they can be entirely divorced from demographic growth. After all, the populations of developed regions like North America and Europe are still growing, albeit at comparatively slow rates. Moreover, the fast-growing populations of the developing countries also

[23] For these quotations see Brown and Kane (1995, p. 154) and Ehrlich *et al.* (1993, p. 18).

contribute increasingly to these high-level environmental threats. Indeed, the discharge of some major greenhouse gases like methane (which is released from rice paddies) and nitrous oxide (which is released by nitrogen fertilizers) probably partly reflects increased agricultural exploitation in the developing world due to the rising demand for food. This is just another example of the many complex synergistic interactions which are involved. To sum up:

> A great deal of the environmental destruction caused by *Homo sapiens* . . . is a direct consequence of the struggle to feed a rapidly expanding population. At the same time, environmental damage constitutes an increasingly important constraint on the future expansion of food harvests . . . Humanity is thus confronted with a serious dilemma: a population–environment–food 'trap'.[24]

With, as we have seen (Table 1.1), the world's population having grown by about 841 million people during the 1980s, the neo-Malthusian case proceeds to argue that these various constraints and threats are *already* having extremely serious and detectable effects on food production. The 'green revolution' which kept food output ahead of world population growth during the 1960s and 1970s is portrayed as running out of steam. The best locations for planting the high yielding varieties of major cereal crops like rice and wheat have largely been used up. It is contended that many countries have reached the point where higher and higher doses of fertilizer to each hectare of cropland produce very little increase in yields. In lots of places the yields of major food crops are said to be falling, or only increasing slowly. Lester Brown says that there has been a 'dramatic slowdown', an 'abrupt deceleration' in the rise of the global grain yield. Norman Myers reports that the situation is so bad that 'the world has now experienced several years of "plateauing" in crop yields.'[25]

According to the neo-Malthusian case the 1970s saw increasingly frequent setbacks in the race between population and food. But in the 1980s there occurred a critical turning point. Since about 1984 the world's population has been growing faster than its production of cereal grains (which form the principal basis of the human food

[24] See Ehrlich *et al.* (1993, pp. 20–1).
[25] See Brown (1994a, p. 185) and Myers (1991, p. 31).

supply). Put differently, since the mid-1980s the global output of grain per person has been falling.[26] Furthermore, this negative trend seems to be established in every world region:

> The worldwide rise that started following World War II was reversed first in Africa . . . in 1967 . . . The next region to peak was Eastern Europe and the [former] Soviet Union, where . . . grain production per person has fallen 8 percent since 1978.
>
> Per capita grain production in both Latin and North America peaked in 1981 . . . In the two remaining regions, Western Europe and Asia, production per person peaked in 1984.[27]

These trends mean that the world's reserve stocks of grain have also been falling. Therefore a couple of exceptionally bad global harvests – arguably more likely under conditions of rapid climate change – could easily cause a repeat of the food price rises and 'world food crisis' which occurred in the early 1970s.[28] Beyond this, it is predicted that during the 1990s global agricultural production may grow at an average annual rate of only 0.9 per cent while population continues growing at around 1.6 per cent. This implies that per capita food production could fall by 7 per cent in the coming decade and raises the prospect of 'an absolute global shortage of food' leading to widespread starvation and famine in developing countries.[29]

Thus the neo-Malthusian view is that in the decades ahead food shortages will increase in both frequency and severity. The next ten years could see a major crisis involving 'mounting grain deficits, surging grain prices and spreading hunger among ever larger numbers of people'.[30] It is seriously proposed that by 2020 there could be several hundred million excess deaths stemming from hunger and famine.[31]

[26] This fact has been very widely cited in the neo-Malthusian and related literature. See, for example, Brown (1993, p. 11), Brown and Kane (1995, p. 38), Ehrlich *et al.* (1993, p. 5), Myers (1991, p. 7), Population Action International (1992, p. 2) and Postel (1992, p. 51). See also Harrison (1992, p. 44).

[27] See Brown (1991, pp. 13–14).

[28] See Brown (1988, 1991) and Brown and Kane (1995, pp. 42–3).

[29] The term 'prediction' in this context is used by Ehrlich *et al.* (1993, p. 23). See also Brown (1988).

[30] See Myers (1991, p. 34). See also Brown and Kane (1995), Ehrlich and Ehrlich (1990) and Ehrlich *et al.* (1993).

[31] Excess deaths are defined as those which may occur over and above the average experience of the past two decades. See especially Daily and Ehrlich (1990). See also Myers (1991).

A QUESTION OF BALANCE

Clearly, this is an extremely pessimistic view of the future. Indeed it is a view which is sometimes pushed even further.[32] As the minus sign in parenthesis in Figure 1.2 shows, the modern neo-Malthusian case often seems to hark back fairly directly to the simple argument of the *First Essay* in which the death rate rises to check population growth. But there is the added sting that these adverse food trends are said to be happening now, and the expected deaths from starvation and famine are possibly just around the corner.

Some of these neo-Malthusian views are so dire that we can probably safely reject them out of hand.[33] I certainly do not envisage that hundreds of millions of *extra* deaths (i.e. deaths additional to those assumed in mainstream world population projections) will occur in the next few decades because of hunger and famine. Faced with such an extreme suggestion most analysts would probably regard themselves as optimists! Furthermore, additional reassurance on this particular issue can be gained from the fact that several of today's prominent neo-Malthusian writers have been having premonitions of imminent starvation and famine for two or three decades now – on a scale which simply has not

[32] For example, outright scepticism is expressed about the validity of population projections for some countries in Africa and Asia because they do not assume that major rises in death rates will occur. Contrary to such population projections, Ehrlich and Ehrlich (1990, pp. 16–17) state that 'human numbers are on a collision course with massive famines anyway'. Much of humanity today is portrayed as living in a demographic 'trap' in which, in essence, population growth creates the conditions which will limit population growth (see Brown and Kane 1995, pp. 55–6). Perhaps most alarmingly, the notion that some populations may be 'trapped', and the bald proposition that the world's population has been outgrowing its production of cereal grains since 1984, are used as evidence to question whether public health measures to reduce death rates should be introduced into poor countries if these measures thereby accelerate the rate of population growth; see King (1990, 1991, 1992).

[33] This book is not much concerned with the arguments summarized in footnote 32. Back in the 1950s various demographic 'traps' were posited which, it was suggested, might prevent developing countries from reducing their death and birth rates. For example see Nelson (1956) and Leibenstein (1957). Subsequent experience has cast grave doubt as to the existence of any meaningful 'traps'; as Preston (1975, p. 241) has eloquently stated of one such trap: it 'shuts so slowly that escape seems inevitable'. For many objections to the suggestion that public health measures be deliberately withheld from some poor populations see, for example, the various contributions to Hammarskjöld *et al.* (1992).

occurred.[34] A charitable interpretation might be that publicizing such premonitions often forces people to wake up and take preventive action. But other interpretations are possible.[35]

This said, some features of the neo-Malthusian argument must be taken more seriously. It would be absurd not to acknowledge that population growth may lead to reduced per capita supplies of cropland and renewable water in the years ahead. It is clear that in many parts of the world rapid demographic growth is contributing to damage the environment, and that this often has adverse consequences for local food production. Furthermore, no one doubts that there will be major growth in global food demand in the next few decades – largely because of population growth. This will certainly present humanity with a considerable challenge.

Moreover, certain elements of the neo-Malthusian case can be evaluated comparatively precisely. For example, since the mid-1980s either the world's population has been growing faster than its cereal production, or it has not. Similarly, either population growth has been outstripping growth of cereal production in all the world's main regions, or it has not. Likewise, either cereal yields show signs of plateauing, or they do not. Again, either there are signs that famines and food crises are becoming more frequent and severe, or there are not. It is true that these issues are not entirely devoid of judgement. But it is also true that they involve relatively 'hard' and objective sets of data. Moreover, demographic studies can provide us with some comparatively firm indications as to the likelihood that starvation and famine will curtail world population growth in the next few decades. By evaluating these rather more specific neo-Malthusian claims we can get a better idea as to

[34] 'The battle to feed all of humanity is over. In the 1970s the world will undergo famines – hundreds of millions of people are going to starve to death in spite of any crash programmes embarked upon now. At this late date nothing can prevent a substantial increase in the world death rate.' See Ehrlich (1968, Prologue). See Table 1.1 for what actually happened to the world death rate in the 1970s and 1980s. In the mid-1970s Brown and Eckholm somewhat less starkly stated that 'The deterioration of the world food situation during the first half of the current decade, together with currently foreseeable trends, also makes it quite clear that the world cannot remain long on its present demographic path. The choice is between famine and family planning, for future population growth clearly will be reduced by rising death rates, as is already occurring in some African and Asian countries, if it is not reduced by declining birth rates' (Brown and Eckholm, 1974, p. 12).

[35] Extreme statements may attract attention. Another perhaps more commendable approach is that of Blaxter (1986, p. 3) who attempts 'to avoid overstatement of the seriousness of our current predicament; it is already serious enough.'

whether the more extreme forecasts deserve any consideration at all.

Of course, in assessing the world's food prospects we must also consider those factors which may boost the total volume of future food production. Several are shown in Figure 1.3. As we shall see, in some parts of the world it is still possible to exploit new supplies of land and water. Then there are the already-noted possibilities that farmers themselves will take steps to improve yields (perhaps partly as a response to population growth) and that the fruits of 'science' will continue to raise agricultural productivity. In addition, enhancements of general infrastructure, better levels of knowledge and education, and improvements in policies (perhaps a perennial hope!) may also advance future food output.[36] We group all of these factors together in Figure 1.3 because – particularly in today's increasingly integrated world – they often combine and interact. Thus a local innovation in land use made by farmers in one location may occur with respect to an entirely new crop, introduced by the national extension services, but developed by an agricultural research centre situated on the other side of the world.

It is clear from Figure 1.3 that in addressing the determinants of food production in the modern world we are explicitly acknowledging an extremely complex contemporary interface between population and food. It is apparent too that this complexity involves possibilities for increased agricultural specialization and trade. Furthermore, whereas neo-Malthusians tend to focus on the death rate as the primary mechanism whereby some kind of equilibrium between population and food may be maintained, it is obvious that this is far from being the sole possibility.

We have already hinted that in many locations migration is an early demographic response to mounting population pressure. But we have also seen that the world's birth rate is falling, and that therefore the global rate of population growth is now coming down (see Table 1.1). The reasons for this birth rate decline are certainly very complicated and vary from place to place. It is highly improbable that many people in the developing world are now turning to contraception and limiting the size of their families simply because they are short of food. But, combined with gen-

[36] One must surely agree with Nathan Keyfitz (and Ehrlich *et al.* (1993, p. 22) when they quote him) that 'If we have one point of empirically backed knowledge, it is that bad policies are widespread and persistent. Social science has to take account of them.' See Keyfitz (1991, p. 15).

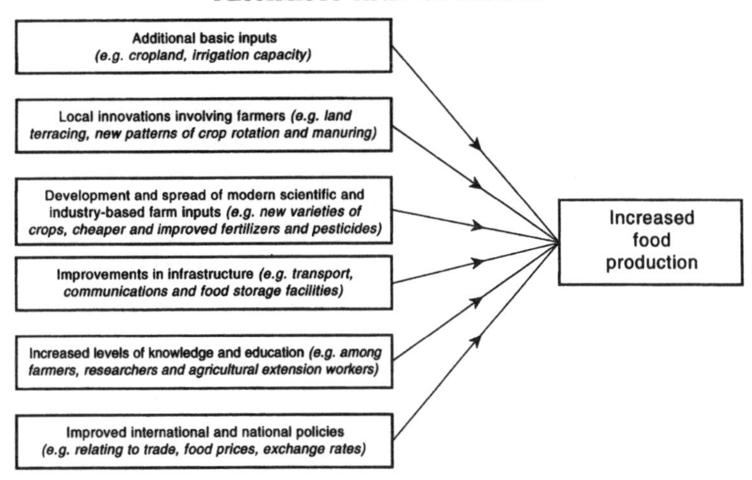

Figure 1.3 Factors which may increase future food production

erally rising expectations about the kinds of lifestyles – including levels and patterns of food consumption – which people may reasonably expect to experience, it seems entirely plausible to regard the falling global birth rate as in part a response to increasing pressures from population growth in a world of constrained resources.[37] So contemporary population growth is probably best viewed as stimulating a complex array of demographic, agricultural, economic and indeed other kinds of adaptive response.

Figure 1.3 makes no specific mention of population growth. That is, it does not address the issue of whether population growth *per se* has a positive or negative effect on per capita food production. To accept that the volume of world food output probably can be increased obviously does not necessarily mean that one accepts that this will happen faster than population growth. It is certainly difficult to agree with any extremely optimistic perspective which holds that the coming increase in world population can be relied upon *by itself* to provoke higher per capita food supplies and reduced chances of famine. In this context it is obviously significant that – rather than simply relying upon market mechanisms and the inventiveness of farmers and agricultural

[37] In a valuable treatment of such issues Bilsborrow (1987) suggests, for example, that China's very rapid birth rate decline in the 1970s must be seen in the context of its very limited supplies of agricultural land and the country's (then) very strict government restrictions on out-migration from rural areas.

scientists working for private companies – today many national governments and international agencies are major sponsors of agricultural research. Interestingly, the basis of the anti-Malthusian case since at least the Second World War has rarely been marked by much *laissez-faire* in this respect. Instead, those writers who have quite correctly argued that food production was likely to keep abreast of world population growth have themselves accepted that this would be a huge challenge, which would probably entail a constant process of adaptive, incremental and evolutionary change.[38]

The abstract question of whether demographic growth by itself raises or lowers levels of per capita food production is probably unanswerable. But in practice we can be sure that the answer is likely to depend upon specific circumstances. Given the very considerable – and largely inevitable – global demographic growth which will occur in the next few decades it is certainly important to address the issue of how, and how well, the world's people are going to be fed. It is really incumbent upon us to regard this future population growth as a challenge from the viewpoint of food production. *A priori* we might expect that food prospects will be mixed – bright in some places and difficult in others. Clearly, our task in this book is going to involve some geographic division of the world into regions. But in an increasingly integrated global environment and economy, the total picture is likely to be greater than merely the sum of the individual parts.

CONCLUDING REMARKS

The clash between the extremes of pessimism and optimism probably says as much about the nature of the human mind as it does about food prospects in the modern world. This also helps to

[38] Notable examples of this perhaps more realistic brand of optimism were Sir E. John Russell and the eminent American agricultural economist Merrill K. Bennett. See Russell (1949) and Bennett (1949/1992). See also Salter (1948) and Woodward (1950). Of course, one might subsume almost any modern changes which have increased food production as ultimately being due to population growth. But such an argument runs the risk of tautology. It seems sensible to draw some distinction between, on the one hand, changes to a particular agricultural system that are in some sense endogenous and unplanned, and on the other hand, changes that are exogenous and planned. While this is only a relative distinction, many scientific and government-sponsored increases in food production clearly fall more in the latter category.

explain the very long history of the basic debate. One might even suggest that the psychology of the dispute has flavoured some of the terms which are commonly used to describe current demographic and agricultural trends – hence the 'population explosion' (pessimism) and the 'green revolution' (optimism).

Certainly, little love is lost between leading protagonists in the contemporary debate. Both sides can be accused of looking past the other.[39] As someone once said:

> In this unamicable contest the cause of truth cannot but suffer. The really good arguments on each side of the question are not allowed . . . their proper weight. Each pursues his own theory, little solicitous to correct or improve it by an attention to what is advanced by his opponents.[40]

If we are going to assess the world's food prospects to the year 2020 then matters of economy and environment need to be considered – and much else besides. It is hoped that the study which follows draws something from both sides of the debate, while not becoming waylaid by either extreme.

[39] For instances of seeming mutual irritation see, for example, Ehrlich and Ehrlich (1990, p. 20) and Simon (1981, p. 71). Environmental issues tend to be highlighted by 'pessimists' and down-played by 'optimists'. The latter tend to be economists, and the former ecologists, environmentalists and kindred spirits.

[40] Ironically the quotation is from the *First Essay*, where Malthus was at his polemical height. See Malthus (1798, p. 5).

2

POPULATION AND FOOD TODAY

INTRODUCTION

This chapter considers several key issues relating to population, food and hunger. The topics addressed include the size and distribution of the contemporary 'world food problem', differences in levels of food availability, food production, trade and aid between the main world regions, and international differentials in food security. We also examine the relationship between current levels of national food availability and rates of population growth. The aim of the chapter is to help set the scene: to provide essential background for our subsequent consideration of past trends and future prospects.

However, we begin by addressing two important preliminary matters, namely (i) the sources and reliability of the data that are to be used, and (ii) the issue of how we will subdivide the world into regions.

DATA SOURCES AND DATA QUALITY

The main types of information used in this study can be broadly categorized as 'demographic', 'agricultural' and 'environmental'. The fact that these data are frequently presented in quantitative form often gives a misleading impression of their accuracy. It is important to appreciate that virtually all of the statistics we will examine are estimates – each with their own particular problems of definition and measurement. We cannot discuss all these problems here. But a few words of explanation – and caution – are in order.

Our main source of demographic data is the latest United

Nations (UN) revision of world population estimates.[1] This is based upon the results of many different national population censuses and surveys. The revision provides fairly accurate estimates of population totals and population growth rates for the world's countries from 1950 onwards, together with corresponding estimates of birth rates and death rates. All these demographic statistics have problems. Indeed, for a few countries even the national population totals may have sizeable error-margins attached.[2] Nevertheless, these demographic statistics are certainly the most trustworthy set of data which we will use.

For many reasons data relating to agriculture and food tend to be much more difficult to interpret. For example, statistics on agricultural output often fluctuate greatly from year to year – reflecting influences such as changes in crop prices or weather conditions. The seemingly simple category 'food' is in reality extremely complex. It contains such disparate items as nuts, pulses, fruits, cereals, vegetables, sugars, roots, edible oils, milk and meat. Moreover, these foodstuffs can be measured in entirely different ways – for instance, by their volume (i.e. weight), or by either their food energy (i.e. calorie) or monetary value.

The sources of data on agriculture and food on which we mostly draw in this book are those of the UN Food and Agriculture Organization (FAO). Each year FAO publishes a huge quantity of extremely detailed statistics on agricultural production, consumption, trade, aid, and much else besides. These statistics are published for virtually all countries, and they relate to a multitude of different foodstuffs.[3]

However, this very profusion and detail immediately raises suspicions about data quality and accuracy. Systems of agricultural data collection are deficient to some degree in all countries. In many developing countries such systems are practically nonexistent. So it

[1] See United Nations (1994).

[2] A quick illustration: the UN revision of world population estimates conducted in 1988 put the 1980 population of Guinea at 5.4 million (see United Nations 1989). But the 1994 revision puts Guinea's 1980 population at 4.5 million, a downward adjustment of 17 per cent.

[3] For example, one major FAO volume used here (see FAO 1987a) includes, either singly or in aggregate form, annual statistics for the period 1948–85 on the production of over 235 agricultural commodities relating to 170 countries. Other annual FAO statistical publications on which we draw either directly or indirectly are the *Production Yearbook*, the *Trade Yearbook*, *Food Aid in Figures* and the *Fertilizer Yearbook* (all published in Rome by FAO).

is not surprising that FAO implicitly seems to accept that a large number of its own published statistics are essentially just educated guesses.[4]

But, while recognizing such problems, perhaps we should not dwell too long upon the rather patchy basis of contemporary agricultural and food statistics. For one thing, such data often benefit from aggregation. That is, we can usually place greater confidence in conclusions when they are based upon statistics for several similar crops, several years' harvests, or when data for several countries are combined to give regional estimates. It must be said that a striking feature of the general literature on agriculture and food is the way in which different authors reach different conclusions, or support different arguments, because of minor differences in the particular time periods they choose to select. Accordingly, some form of aggregation, or averaging, is adopted throughout this book.[5]

Another way in which this study tries to minimize problems relating to food and agricultural statistics is by focusing chiefly upon cereals (i.e. wheat, rice and coarse grains). Wheat is the world's most important single cereal crop measured in terms of the tonnage which is produced. Rice is the world's most important single crop measured in terms of the number of people – mostly in Asia – for whom it is their staple food. The 'coarse grains' category – consisting of barley, millet, oats, rye, sorghum and, most importantly, maize (i.e. 'corn' in American parlance) – collectively exceeds the production of either wheat or rice. Although coarse grains are sometimes grown as food, for example in parts of Asia

[4] On the deficiencies of agricultural data collection systems and statistics see Grigg (1993) and the contributions to Braun and Puetz (1993). The FAO *Production Yearbook* routinely carries the disclaimer that 'As in past years, this volume also includes estimates made by FAO on area and production of major crops . . . where no official or semi-official figures were available from the countries themselves. The publication of these estimates gives the countries concerned a chance to examine them, and it is hoped that they will provide FAO with more reliable figures.' See FAO (1994, p. vii).

[5] Also, full time series data (rather than statistics for selected periods) are presented whenever possible. The issue of how time series can be interpreted in selective ways is touched on in chapter 3.

and Africa, in the rest of the world they tend to be used predominantly as feed for livestock.[6]

In the early 1990s nearly half of the world's total area of cropland was devoted to the production of cereal crops. A similar proportion of total human food energy (i.e. calorie) intake is accounted for by the *direct* consumption of cereals – in such forms as cooked rice, bread and porridge. And if allowance is made for *indirect* consumption – through the large quantities of cereals which are fed to livestock and eventually consumed as meat – then around two-thirds of total human calorie intake is accounted for by cereals.[7] Concentrating upon cereal crops has the advantage that the quantities produced can be decomposed into the area of cropland that is harvested of cereals and the level of the cereal yield per unit of area.[8] Finally, the fact that cereal cultivation occupies such a central place in agriculture – and indeed life – in many countries probably means that statistics on cereals are somewhat more accurate than those relating to many other crops.[9]

Lastly in this section, a brief word is in order regarding data on the environmental conditions affecting world food production. We might characterize the quality of the available population statistics as 'pretty good'; and we might describe the quality of the information on agricultural production as 'variable, but probably OK for present purposes'. But the coverage and quality of most relevant environmental data are much worse. To give just one illustration,

[6] Maize accounts for about 60 per cent of total world coarse grain production. It is the dominant cereal in the world grain trade, though this is primarily for livestock feed; wheat is the world's principal traded grain for food; rice is not traded much internationally. Wheat and rice are overwhelmingly used as food. Important non-cereal crops, in terms of tonnage produced and area harvested, are potatoes and soybeans. The main potato-producing countries are Russia, Poland, China, Ukraine, the United States and India. Soybean production is dwarfed by the output of the main cereal crops (i.e. wheat, rice and maize) and is largely concentrated in the United States, Brazil, Argentina and China.

[7] Direct consumption of cereals is the most important single element in most national diets. On the composition of total human food energy intake see Alexandratos (1988).

[8] In this study all yields are expressed in terms of metric tons (mt) of output per hectare (ha) of harvested area. In ascertaining harvested area, a hectare of land is counted each time it is cropped in a year.

[9] The United States Department of Agriculture (USDA) provides an alternative set of estimates of cereal production and trade to those of FAO, but for a smaller number of countries. Broad indications on trends from both sets of estimates are similar at the world or world regional levels. But significant differences as to cereal volumes produced or traded often arise at the country level. See Paulino and Tseng (1980) and Ravi Kanbur (1991).

the evidence available to assess the extent of global land degradation and its consequences for agricultural production has rightly been described as 'skimpy and anecdotal'.[10] The problems with the environmental data-base arise partly because the enterprise of systematically gathering information on environmental conditions is often rather new. Also, the sheer technical and logistical difficulties of assembling global data sets on, for example, the extent of soil erosion or the depletion of water supplies are very great. Moreover, an especially large element of judgement may be involved in assessing changes in environmental conditions – for example, as to whether or not the Sahara desert is creeping southwards. Therefore, whether we consider the extent of world soil erosion, water supplies and their depletion, or for that matter the existence of and threat posed by global warming, the scope for uncertainty and disagreement over environmental trends tends to be particularly great.[11]

DIVIDING UP THE WORLD

Attempting any global assessment immediately raises questions as to how the world should be divided up. There are obvious limitations in treating humanity as a single unit. Indeed, one of the main themes of this book is that concentrating upon global measures can give a very misleading view of food production trends. But, on the other hand, it is clear that we cannot examine every country individually. Accordingly, the approach used here is to conduct the analysis mainly in terms of major world regions – but to pepper the discussion with mention of countries (especially those with larger populations) where appropriate.

The seven world regions on which this study is based (and the colloquial names for them which we will use) are listed in Table 2.1. They derive directly from the standard United Nations regional classification of countries which was holding in 1992, and which groups together the nations of the former Soviet Union (FSU). The regions are broadly homogeneous in terms of their demographic, cultural, economic and agricultural characteristics –

[10] See Crosson and Anderson (1992, p. 42). See also Pearce (1992).

[11] The principal environmental data used in this work are the compilations of the World Resources Institute and the Worldwatch Institute. See Brown *et al.* (1993, 1994) and World Resources Institute (1990, 1992, 1994).

although inevitably there is also very significant variation within each region.[12]

Sub-Saharan Africa consists of all countries in Africa except those in northern Africa. So, as defined here, sub-Saharan Africa includes South Africa. This region contained about 9 per cent of the world's population in 1990 (see Table 2.1). The Middle East refers collectively to all countries in northern Africa and western Asia, stretching from the Arabian peninsular in the east to Morocco in the west. This region includes both Sudan and Turkey. Demographically the Middle East is the smallest of our seven regions, containing only about 5 per cent of humanity in 1990. The next region is South Asia, which contains nearly a quarter of humankind. It comprises the countries of the Indian subcontinent, plus Iran. India is the most important of these countries – with about 71 per cent of the region's total population. The Far East is demographically the largest of our regional groupings and it is also one of the most diverse. This region covers all countries in eastern Asia (including Japan) and south-eastern Asia. China alone contains one-fifth of humanity – 64 per cent of the total population of the Far East. The next region, Latin America, comprises all countries in the Caribbean and South and Central America, including Mexico. In 1990 Latin America contained about 8 per cent of the world's population.

The two last regions in Table 2.1 are also economically the most advanced. All countries in Europe and the former Soviet Union have been grouped together to form the third largest regional population, with about 15 per cent of humanity. Finally, North America and Oceania have been combined into a single hybrid 'region' – because the United States and Canada (which together comprise North America) share many relevant characteristics with Australia (the predominant country of Oceania).

Within these seven regions all countries which in 1990 had populations of 5 million or more have been selected for particular attention. These 91 'sample' countries are listed in the Appendix together with indicators of their demographic, socio-economic,

[12] Intra-regional issues are considered mainly in chapter 6. For a listing of countries by the UN regional classification used here, see United Nations (1993).

Table 2.1 Population of main world regions in 1990 and measures of regional representativeness of sample countries

Region	Total population (million)	Total population (%)	Number of sample countries	Population of sample countries (million)	% of regional population in sample countries	% of regional cereal production produced in sample countries
Sub-Saharan Africa	490	9.3	25	453	92.4	95.7
Middle East	276	5.2	10	252	91.3	98.5
South Asia	1193	22.6	7	1191	99.8	100.0
Far East	1794	33.9	11	1778	99.1	99.6
Latin America	440	8.3	15	413	93.9	95.6
Europe/FSU	788	14.9	20	765	97.1	98.4
North America/Oceania	304	5.8	3	295	97.0	99.8
World	5285	100.0	91	5147	97.4	98.9

Notes: (a) Here and in all subsequent tables totals do not always add due to rounding

(b) FSU denotes the former Soviet Union

(c) Measures of cereal production relate to average output for 1989–91

(d) In general in subsequent tables regional indices are weighted averages for the corresponding sample countries within each region. The principal exception relates to total quantities for regions, where relevant totals have usually either been calculated directly or obtained by proration

Principal sources: United Nations (1993, 1994)

food and agricultural status.[13] Table 2.1 shows that in 1990 the combined population of the sample countries constituted over 97 per cent of humanity. And it is certain that the sample accounts for a similarly high fraction of global food consumption. Note too that the sample countries account for the bulk of the populations and food output (as gauged here by cereal production) of each of the seven world regions. Finally, the sample consists of 24 developed countries (i.e. all those in Europe/FSU, North America/Oceania, plus Japan) and 67 developing countries (see Table 2.1).[14]

Selected demographic and socio-economic measures for the regions are given in Table 2.2. The population of sub-Saharan Africa is exceptionally poor and growing very fast. If sustained, a growth rate of 3 per cent per year leads to a population doubling in size in about 23 years. This remarkably high rate of demographic increase results from a very high regional birth rate and a much lower – though still relatively high – death rate. In fact, birth rates are declining in a growing number of countries in this region – notably South Africa, but also Kenya, Botswana, Zimbabwe and several others too.[15] But, on average, women in sub-Saharan Africa still have about six live births each, and the prevailing life expectancy at birth is only slightly above 50 years.

Between 1985 and 1995 the populations of the Middle East, Latin America and South Asia (the last, another exceptionally poor region) all grew at 2 per cent per year or more. The growth rate for the Middle East was particularly high. In the mid-1990s birth rates – and therefore population growth rates – in these three regions are generally declining. But there remains considerable scope for further birth rate reduction (especially in large parts of the Middle East and South Asia) and there is also significant scope to raise life expectancy and reduce death rates. Note the higher levels of

[13] Much of the book's discussion – perhaps especially in chapter 6 – involves implicit reference to material in the Appendix. The country statistics given there include population size in 1990, projected percentage population increase for the period 1990–2020, per capita daily calorie supply, GNP per capita, and a measure of national self-sufficiency in cereals.

[14] The conventional distinction between developed and developing countries is used here, although some countries are clearly jumping the gap. The exclusion from the sample of small city-states like Kuwait, Singapore, and the non-nation of Hong Kong, which all have minor agricultural sectors, helps to explain why in Table 2.1 the sample countries account for slightly lower proportions of population than of cereal production in each region. The situation of new countries formed from the former Soviet Union is considered briefly in chapter 6.

[15] See, for example, Foote *et al.* (1993).

Table 2.2 Selected demographic and socio-economic measures, world regions, around 1990

Region	*Population growth rate 1985–95 (% per year)*	*Crude birth rate 1985–95*	*Crude death rate 1985–95*	*Total fertility rate, 1985–95 (per woman)*	*Life expectancy at birth, 1985–95 (years)*	*Urbanization 1990 (%)*	*GNP per capita 1990 ($US)*
Sub-Saharan Africa	3.0	45.6	15.6	6.4	50.7	31.0	480
Middle East	2.6	33.8	8.7	4.7	62.7	53.4	1536
South Asia	2.2	32.7	10.7	4.4	59.0	27.3	444
Far East	1.4	21.6	7.3	2.4	67.5	37.0	2380
Latin America	2.0	27.0	7.0	3.2	67.7	71.5	2148
Europe/FSU	0.4	14.2	10.3	1.9	73.0	70.7	11,719
North America/Oceania	1.1	15.8	8.5	2.0	75.7	74.8	21,391
World	1.7	26.0	9.5	3.2	63.7	45.2	4180

Notes: (a) The crude birth and death rates are expressed per 1000 population per year

(b) The total fertility rate is the number of live births an average woman would have by the end of her reproductive years on the assumption that prevailing rates of age-specific fertility remain constant

(c) Because of variation in the definition of 'urban areas' between countries the urbanization statistics should be regarded as especially approximate

Principal sources: United Nations (1991, 1993, 1994); World Bank (1992); World Resources Institute (1992)

average incomes and urbanization found in the Middle East and, still more, Latin America (see Table 2.2).

By most criteria the Far East is more advanced than the regions considered so far – although this partly reflects the presence of Japan. Around 1990 the average level of fertility in the Far East was about 2.4 live births per woman and life expectancy was around 68 years. Both these regional characteristics largely reflect the success of China in reducing its birth and death rates. Indeed, the dramatic reduction of the Chinese birth rate chiefly explains why this region's population grew at the comparatively slow rate of 1.4 per cent per year during the period 1985–95.

Finally, the most developed regions – Europe/FSU and North America/Oceania – experience low birth and death rates. On average, women in these regions have about two live births each and life expectancies are well over 70 years. Therefore population growth rates are relatively low. Indeed, especially in the case of North America/Oceania in-migration from other regions is a significant cause of population growth (though still secondary to natural increase). As one would expect, average levels of income and urbanization in these regions are high (see Table 2.2).

With this as background, we now consider the contemporary world food situation, beginning with a consideration of the extent of human hunger.

THE DEFINITION AND EXTENT OF HUNGER

There is widespread, if not universal, agreement that today several hundred million people probably don't have enough food to eat – and that most of these people live in Africa and Asia.

However, in trying to go beyond this simple statement the problem immediately arises as to the *criteria* to be used in defining people as inadequately fed. This problem was illustrated by the so-called 'great protein fiasco' of the 1960s and 1970s when a UN expert committee set unrealistically high standards to define adequate protein intake. The subsequent amendment of these standards led to a significant downward revision of the number of people considered to be suffering from insufficient protein consumption.[16] The general view which now prevails is that most diets which provide sufficient food energy – i.e. calorie – intake probably also provide enough protein, although young growing

[16] See Blaxter (1986) and Grigg (1993, pp. 7–9).

children, and pregnant and lactating women, may have additional protein requirements. Also some traditional diets based upon root and tuber crops (e.g. cassava, potatoes, yams) may provide enough food energy, but not enough protein.[17]

However, to reiterate, the world food problem today is primarily defined in terms of dietary energy deficiency, or ***undernutrition***. This constitutes the principal criterion for a modern definition of 'hunger'.[18] We have already seen that cereals are the most important single element in the human diet. And in this context it is significant that, in addition to their calorie content, all cereals contain significant quantities of protein, albeit of variable quality.[19]

Only a few countries conduct systematic and nationally representative surveys of food consumption which enable direct assessments of calorie intake to be made. So the most widely cited estimates of food energy intake and the prevalence of undernutrition are produced by FAO using various indirect procedures.[20] For each country estimates are made of the total quantity of food available for human consumption in a given period. This involves assembling data on national food production, trade and aid, and then making deductions for quantities of food which are lost in storage, fed to animals or used as seed. Next food composition factors are used to convert the food availability estimates into calorie equivalents. And, using population totals, these are expressed as national estimates of per capita daily calorie supply.[21]

[17] Diets based upon root and tuber crops are found in many specific locations, for example parts of the Pacific and the Andes in South America. But the main areas where such crops still seem to predominate in national diets are in central and west Africa.

[18] Of course, quantities of vitamins and minerals are required for an adequate diet. Thus one FAO definition of malnutrition is 'a pathological state, general or specific, resulting from a relative or absolute deficiency or an excess in the diet of one or more essential nutrients'. See FAO (1987b, p. 70). Clearly, this definition can pertain to calories, proteins, vitamins and minerals. In this book our concern is more with ***undernutrition***. But in the developing world many people also suffer from inadequate intake of specific nutrients. See FAO (1987b), FAO/WHO (1992) and Kutzner (1991).

[19] See Kutzner (1991, pp. 170–3).

[20] These procedures are integral to FAO's periodic 'world food surveys' which, however, are not surveys in the conventional social-scientific sense. For further details see, for example, FAO (1987b, 1992).

[21] FAO stresses that such per capita calorie supply estimates relate to the quantity of food reaching the consumer. The quantities actually consumed may be lower, due to further losses incurred in the household (e.g. in storage, preparation or cooking); alternatively, the amount of food consumed may be higher if, as is the case in some poor countries, wild plants and wild meats – which do not appear in the FAO statistics – are significant elements in the diet.

Because food is never distributed equally between all households in a country – the rich typically getting more than average, and the poor typically getting less – FAO then attempts to estimate the distribution of calorie supplies within each national population, using various statistical procedures and assumptions, and whatever information on patterns of household income or expenditure may be at hand.

But perhaps the most contentious step in FAO's process of gauging the extent of undernutrition is that of estimating national per capita calorie *requirements*. This involves estimating the basal metabolic rate (BMR) for each country – i.e. the number of calories required for an average person to maintain essential body functions, while lying at complete rest – taking account of variation in factors like national bodyweights and population age composition. Levels of calorie requirements are obtained by multiplying the BMRs by a constant. The value of this constant has been repeatedly revised upwards with time, which, other things being equal, *increases* the number of people estimated as undernourished. The value used recently is 1.54 – a factor judged to be sufficient to allow people to engage in 'minimal activity for productive work'.[22] Finally, comparison of the estimated calorie requirements with the estimated levels and distributions of calorie supplies provides estimates of the *numbers* of undernourished people.

Clearly, such a complicated and indirect approach, embodying a host of assumptions, and ultimately being based upon rather patchy empirical foundations, is far from ideal! Quite minor changes to the methods and assumptions can produce very large differences to the estimated numbers of undernourished people.[23] And it is not inconceivable that an organization which is devoted to food and agricultural production may have an interest in presenting estimates of global hunger which err on the high side.

Moreover, a fundamental challenge to FAO's whole approach stems from the so-called 'Sukhatme–Margen' hypothesis which

[22] See FAO (1992, p. 8). Previous FAO assessments have used multiplying constants of 1.2 and 1.4; see FAO (1987b).

[23] As Kellman (1987, p. 28) states: 'The line below which misery is defined has been quite sketchily drawn, while the tool measuring food supplies is clearly imperfect.' Or to quote Kates (1993): 'No one really knows how many hungry people there are in the world.'

Table 2.3 Measures of undernutrition and calorie supply, world regions, around 1990

Region	*Number undernourished (million)*	*% of population undernourished*	*Daily per capita calorie supply*	*% of calories obtained from vegetable products*	*Daily calorie supply as % of requirements*
Sub-Saharan Africa	168	33	2250	93.4	93.7
Middle East	31	12	3018	90.6	119.7
South Asia	} 528	} 18	2215	92.9	98.7
Far East			2638	79.3	113.1
Latin America	59	13	2699	82.5	115.3
Europe/FSU	—	—	3422	69.1	132.6
North America/Oceania	—	—	3586	69.1	136.2
World	786	15	2697	84.3	112.7

Notes: (a) All figures refer to 1988–90

(b) Estimates of numbers undernourished (strictly, those defined as 'chronically undernourished') are for slightly different geographical regions to those shown and exclude countries of under 1 million inhabitants. Also, these estimates are unavailable for sub-regions within Asia. Remaining estimates are based on sample countries

Principal sources: FAO (1992, 1993); World Resources Institute (1992)

contends that populations can adapt to low food energy intakes without much changing their levels of physical activity, their bodyweights, or even impairing their health.[24] On the other hand, it can be argued that the levels of calorie requirements set for many poor countries are actually too low, because they make no allowance for the adverse health conditions which people commonly face. Heavy loads of infectious disease in the environment can be just as responsible for inadequate dietary intakes as are low levels of food energy availability *per se*. And, in such conditions, a strong case can be made for increasing the level of calorie supply requirements (which would raise the estimates of numbers of undernourished people).[25]

Despite these significant problems, the FAO estimates of undernutrition and calorie supply are the most comprehensive and authoritative which are available. Table 2.3 shows that they suggest that there were about 786 million undernourished people around 1990.[26] This represents roughly 15 per cent of the world's population, or about 20 per cent of the population of the developing world. The greatest number of undernourished people – around two-thirds of the total – live in Asia. But sub-Saharan Africa is the region with the greatest proportion (33 per cent) estimated as undernourished. Both the Middle East and Latin America contain much smaller numbers of hungry people, and the corresponding proportions undernourished are put at around 12–13 per cent. Lastly, while there certainly are undernourished people in the two most developed regions, in comparative terms the numbers are negligible. Instead, the main dietary problems of Europe/FSU and North America/Oceania lie more in overnutrition

[24] See Sukhatme and Margen (1982), and for a useful critique of the hypothesis see Kumar and Stewart (1992).

[25] For example, diarrhoea is perhaps the leading cause of death for the world's children today. It often produces severe deterioration in nutritional status by reducing appetite and increasing physical loss from the intestine. Infectious diseases are a major independent cause of undernutrition. See Tomkins and Watson (1989).

[26] It is worth noting that the World Bank – the only other agency which intermittently produces broadly analogous estimates – generally comes up with higher numbers of undernourished people. For comparisons of FAO and World Bank estimates see Grigg (1993, pp. 27–8), Kates (1993) and Kutzner (1991, pp. 158–9). See also World Bank (1986).

– i.e. excessive calorie consumption, contributing to obesity and increases in various degenerative diseases.[27]

The estimates of daily calorie supply as a per cent of requirements further enhance the picture, although they take no account of inequalities in food distribution (see Table 2.3). Calorie supplies in both North America and Europe are well in excess of requirements. At the aggregate level the same is true of the Middle East, the Far East and Latin America – although because of distributional inequalities these regions still contain appreciable numbers of undernourished people. It is clear from Table 2.3 that within Asia the problem of undernutrition is chiefly concentrated in South Asia. Some writers suggest that the Indian subcontinent is home to at least half the world's hungry people.[28] The FAO estimates indicate that even if South Asia's food supplies were equitably distributed – which is very far from being the case – there would still not be enough to meet requirements. The same is true of sub-Saharan Africa, but to an even worse degree.

Because of their relatively uncomplicated nature the basic daily per capita calorie supply estimates in Table 2.3 have much to commend them – provided they are interpreted with care. These estimates can also be decomposed to indicate the extent to which calorie intake is based on vegetable as opposed to animal products.[29] It can be seen that not only are daily calorie supplies very low in both sub-Saharan Africa and South Asia – at just over 2200 calories per person – but also the diets are overwhelmingly vegetarian. Estimated per capita food energy supplies are significantly higher in the Far East and Latin America, and in both regions animal products account for roughly 20 per cent of total calorie intake. On average, diets in the Middle East contain still more

[27] These problems involve overconsumption of fat, saturated fatty acids, sugar and salt, with inadequate intake of fibre and complex carbohydrates. In recent decades adverse dietary changes in developed countries have been linked to greater consumption of processed foods. There are signs that such detrimental changes are increasingly happening among more affluent, more urban, sections of the populations of developing countries; on the other hand, in some developed countries there are indications of a trend towards more healthy eating patterns. See FAO (1987b), FAO/WHO (1992) and Popkin (1993).

[28] See Kutzner (1991, p. 160); see also World Bank (1992, pp. 29–30). For discussion of distributional issues in the South Asian context see the contributions to Harriss *et al.* (1992).

[29] Although foodstuffs like meat, fish, milk and eggs are not absolutely necessary for a balanced diet, their contribution to total food energy consumption provides a useful insight as to overall dietary variety and quality.

calories – around 3000 per person per day – although their protein content is actually lower than the diets of the Far East and Latin America. Finally, daily calorie availabilities are extremely high in Europe/FSU and North America/Oceania, where animal products account for about 30 per cent of total food energy.

Figure 2.1 combines the data on population and food energy consumption. It shows the 91 sample countries by their estimated daily per capita calorie intakes around 1990, but with their sizes being drawn in proportion to their total populations.[30] The map emphasizes that South Asia and sub-Saharan Africa are the two main sites of contemporary world hunger. The Indian subcontinent combines a very large population and very low calorie supplies; here the situation of Bangladesh is particularly dismal. Countries in eastern and southern parts of sub-Saharan Africa also have especially low calorie availabilities. Major countries acutely affected include Zaïre, Tanzania, Kenya, Ethiopia and – slightly to the north, here classed as part of the Middle East region – Sudan. Other areas where some national populations are poorly provided with food are eastern South America (especially Peru and Bolivia), parts of Central America and the Caribbean, and Southeast Asia.

But, to reiterate, because of distributional inequalities some countries with high overall levels of calorie availability still encompass significant numbers of undernourished people. South Africa is a clear illustration of this, though hopefully the situation there is now beginning to change.[31] Moreover, the sheer size of China's population doubtless also signifies a sizeable number of hungry people, notwithstanding the country's comparatively good record on matters of distribution (at least until recently).

CEREAL PRODUCTION, CONSUMPTION, TRADE AND AID

Estimates of calorie availabilities are convenient, if imperfect, indicators of the amount of food which a country may have at

[30] See the Appendix for the data used in Figure 2.1. For a similar map for 1979–81 see FAO (1987b, p. 12).

[31] According to figures cited by Caldwell and Caldwell (1993a) the per capita income of South Africa's black population in 1990 was probably not above $US 1150, whereas average white incomes were about 15 times higher.

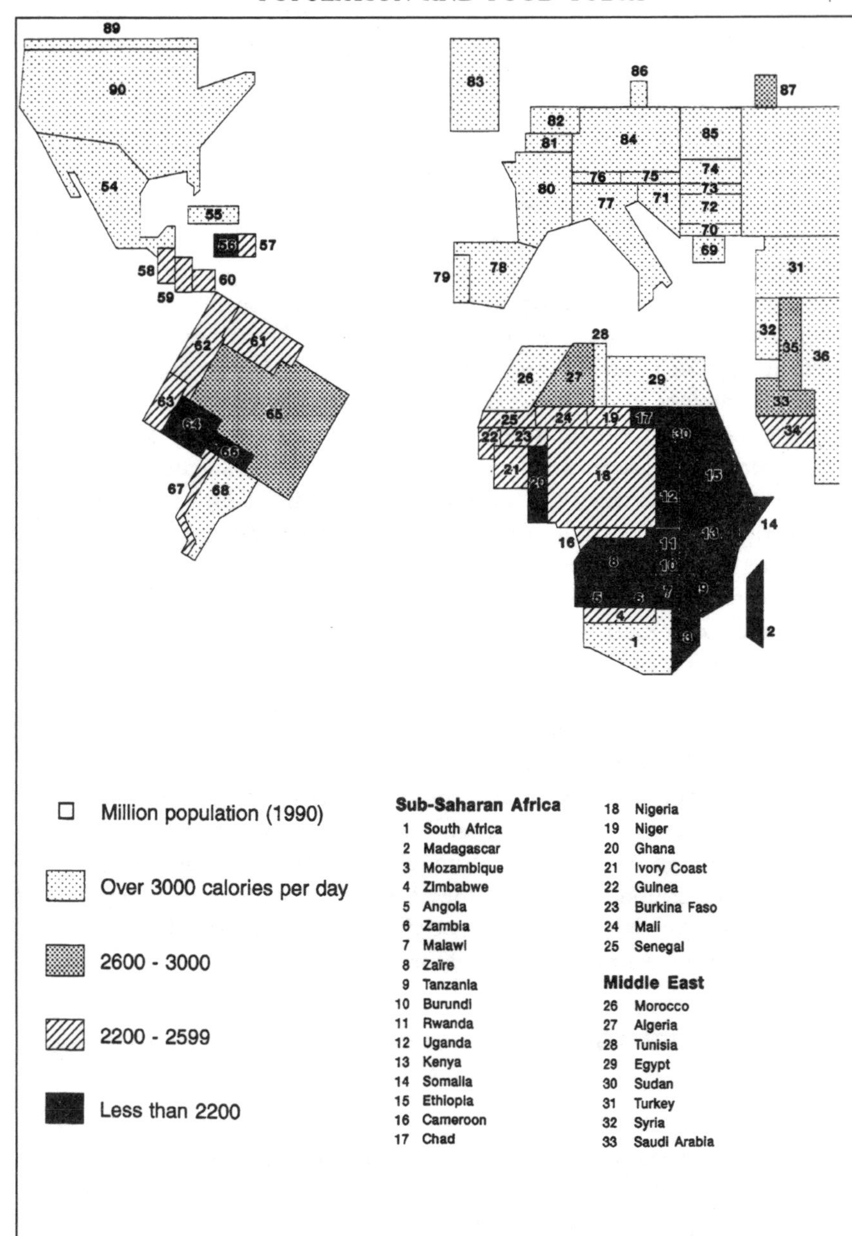

Figure 2.1 Distribution of sample countries by per capita daily calorie supply, around 1990

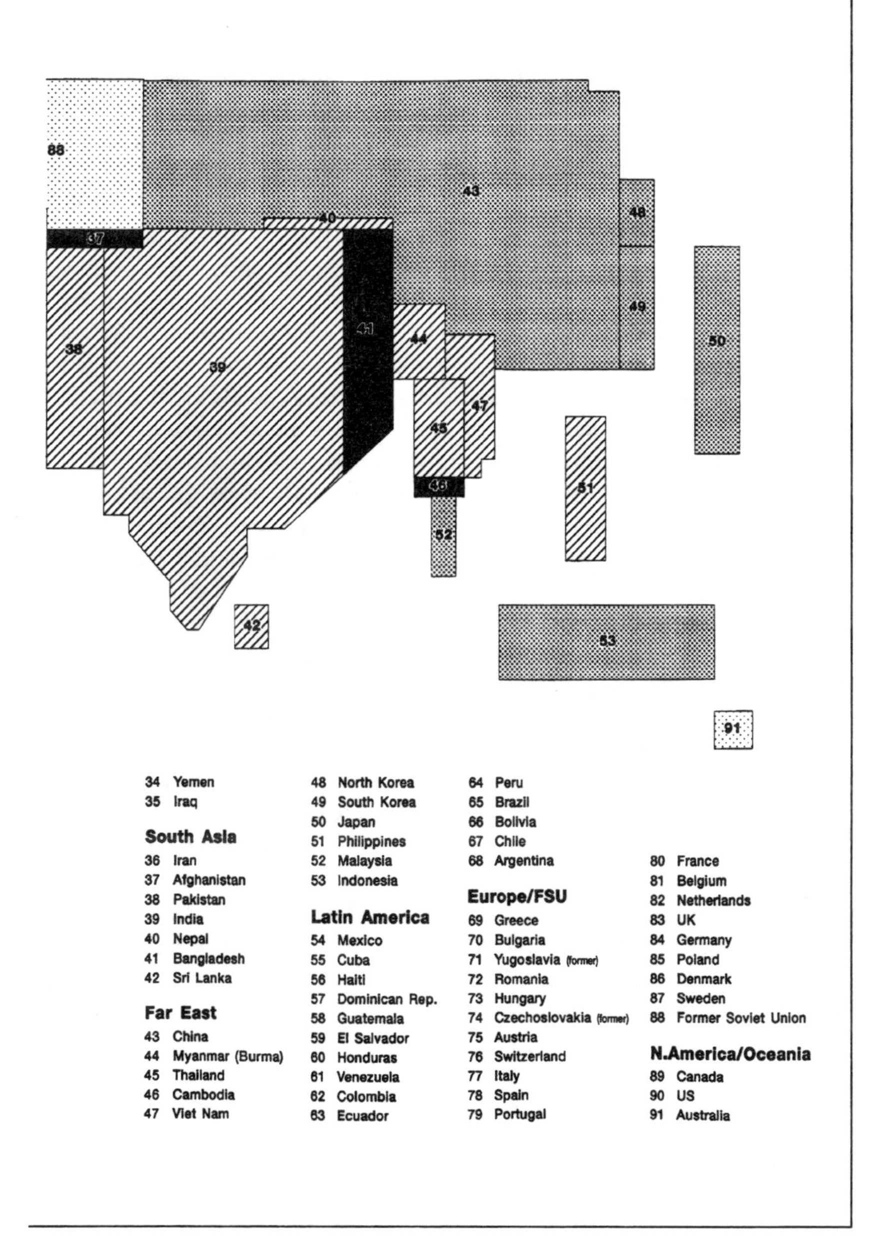

Note: Areas of countries on the map are proportional to the size of population in 1990

hand. But they say little about *how* that food is obtained. Data on cereal production and flows can shed light on this.

Table 2.4 shows that around 1990 global per capita cereal production stood at slightly over 360 kg. But there was tremendous variation between the main world regions. Per capita cereal output is exceptionally low in sub-Saharan Africa, where a variety of cereals – most prominently maize, but also the region's traditional crops of sorghum and millet – are grown. The very low level of cereal production in sub-Saharan Africa is partly explained by widespread reliance upon root and tuber crops – mainly cassava and yams – in much of west and central Africa, where such crops can account for more than half of total calorie intake. However, even in eastern and southern Africa – where cereals form a much more important part of the diet – the average level of cereal production was still only about 160 kg per person around 1990.

Levels of per capita cereal production are also very low in South Asia, where the principal crops are rice and wheat. Production is somewhat higher in both Latin America and the Middle East. But, of the predominantly developing regions, cereal output per person is highest in the Far East. Again, this reflects the comparatively good performance of China where, of course, rice is the principal crop. Finally, both Europe/FSU and North America/Oceania produce very much greater quantities of cereals per person than do all the other world regions. At the extremes, the contrast is between sub-Saharan Africa where per capita cereal production is only about 138 kg, and North America/Oceania where the corresponding figure is over 1.2 metric *tons* (mt) per head (see Table 2.4).

The data on cereal trade illustrate the important fact that all world regions import grain from the traditional exporting bloc of North America/Oceania (Table 2.4). The United States dominates this bloc in terms of both cereal production and exports.[32]

The Far East region imports the greatest volume of cereals, equivalent to about 9 per cent of its own production. By itself Japan is responsible for about half of these imports – other major purchasers being China and newly industrializing countries like South Korea. The second greatest grain importing region is the Middle East, where traded cereal imports are equivalent to a

[32] For example, in the period 1989–91 the US accounted respectively for 80 and 72 per cent of North America/Oceania's cereal production and net traded exports.

Table 2.4 Average annual measures of cereal production, consumption, trade and aid, world regions, around 1990

Region	*Per capita production 1989–91*	*Total annual production 1989–91*	*Net trade 1989–91*	*Donations or receipts 1987–89*	*Net trade as % of production*	*Donations and receipts as % of production*	*% of total consumption used as livestock feed 1990*	*Per capita cereal availability*	
								Total	*Direct*
	(kg)	*(million mt)*	*(million mt)*	*(million mt)*					
Sub-Saharan Africa	138	67.3	3.9	2.7	6	4	8	150	138
Middle East	262	72.2	31.1	3.2	43	4	34	386	255
South Asia	226	269.8	9.7	2.6	4	1	3	237	230
Far East	309	553.6	52.1	0.7	9	0	20	338	270
Latin America	237	104.6	10.2	2.1	10	2	40	265	159
Europe/FSU	619	488.3	11.7	−2.4	2	0	58	634	266
North America/ Oceania	1249	367.7	−129.9	−8.5	35	2	70	780	234
World	363	1923.5	—	—	—	—	38	363	224

Notes: (a) In this and subsequent tables the numbers shown for production, trade and aid should be regarded as broadly indicative rather than exact. There are many reasons for this. For example, the FAO usually repeatedly revises its estimates for recent years – sometimes very considerably – and often in an upward direction. In addition, totals and calculated net figures do not always add, for example due to rounding, differences in time periods, and, not least, divergences in reported statistics between different sources. Similar reasons account for occasional minor discrepancies between different tables

(b) Net exports and cereal aid donations are shown as negative numbers. In their calculation it has been assumed here and subsequently that net trade and aid flows for countries within a region are the result of inter-regional flows. However a small proportion of flows are intra-regional. As is usual in such statistics reported flows do not balance

(c) Data on per cent of consumption used as feed are based on USDA sources. All other estimates are from FAO data

Principal sources: FAO (1993); World Resources Institute (1992)

massive fraction (43 per cent) of total regional production. Most countries in the Middle East are reliant upon imports to a significant extent. Egypt, Algeria and Saudi Arabia together are responsible for over half this region's traded imports. In volume terms the third greatest cereal importing region is Europe. In fact, most countries in Europe are net cereal exporters and the European Union (EU) is actually the world's second largest cereal exporting bloc. However, around 1990 the former Soviet Union was still the world's leading single importer (just ahead of Japan), purchasing cereal volumes roughly equivalent to all traded imports into the entire Middle East.[33]

Table 2.4 shows that in volume terms both Latin America and South Asia are relatively modest cereal importers, with net purchases being equivalent to about 10 and 4 per cent of regional production respectively. Argentina is a major cereal exporter in Latin America, and Cuba, Venezuela, Brazil and Mexico are major importers. Regional imports into South Asia are mainly accounted for by Iran and – to a much smaller extent – Bangladesh. India itself is more or less self-sufficient in cereals. Lastly, note that sub-Saharan Africa is the region with the smallest volume of traded cereal imports – equivalent to only about 6 per cent of its own grain production around 1990.

Cereal aid refers to the transfer of cereals from donor to recipient countries either on highly concessional terms or in the form of a complete grant. Data on cereal donations and receipts must be interpreted with particular care, for as well as being influenced by humanitarian considerations, such flows are motivated by political, economic and other concerns. Donations can change abruptly from year to year – usually for reasons unrelated to changes in the food requirements of recipient populations. However, it is clear from Table 2.4 that modest quantities of cereal aid are received by the countries of Latin America, South Asia, the Middle East and sub-Saharan Africa. And in the last two of these regions donations amount to quite a significant fraction (4 per cent) of total regional cereal production. In contrast, the Far East receives very minor quantities of cereal aid.

Table 2.5 provides an overview of the main world flows of cereal aid. As can be seen, around 1990 the United States alone was

[33] The collapse of the Soviet Union in 1991 has dramatically reduced traded cereal imports into the former Soviet Union. This issue is considered further in chapter 6.

Table 2.5 Average percentage distribution of total global cereal aid flows in 1989–90 and 1990–91, from major donors to recipient regions

Recipient regions	*US*	*EC*	*Major donors* *Canada*	*Japan*	*Others*	*Total*
Sub-Saharan Africa	9	10	2	1	4	26
Middle East	19	5	2	0	1	27
South Asia	9	4	5	1	4	23
Far East	1	1	1	1	1	5
Latin America	15	1	1	0	0	18
Unspecified	0	0	0	0	0	1
Total	54	21	11	4	10	100

Notes: (a) Zero entries sometimes cover small volumes of donations and columns may not always sum exactly due to rounding

(b) Total cereal aid volumes in 1989–90 and 1990–91 according to the sources used here were 10.0 and 12.0 million tons. The latter figure includes approximately 1.3 million tons donated to Poland, Romania and Bulgaria (largely from the US) which have been excluded in the above calculations. As elsewhere numbers shown should be regarded as broadly indicative rather than exact

(c) The 'Others' category mainly reflects donations by Argentina, Australia, Saudi Arabia, several non-EC European countries, plus UN World Food Program purchases

(d) Triangular aid (i.e. where the donor pays for shipments to a recipient country, from a third country) is not generally practised by the United States. But in 1989 it accounted respectively for 32, 7, 54 and 17 per cent of the donations of the EC, Canada, Japan and Others

Principal source: FAO (various years) *Food Aid in Figures*, Rome

responsible for about half of total global donations. The largest quantities of US donations were destined for the Middle East and Latin America. The European Community (EC) was the second largest donor bloc, with Canada in third place. Again, the overall impression is of the predominance of North America as a source of global cereal supplies.

The greatest recipients of cereal aid, by far, are Egypt and Bangladesh. These two countries together received 25 per cent of total world cereal donations during 1989–91. The United States provided about 70 per cent of Egypt's cereal aid, and the US is also the largest single donor to Bangladesh. Clearly, the circumstances of these two recipient countries are quite different;[34] it is reason-

[34] For example, Egypt is a key actor in the Middle East, whereas in South Asia Bangladesh is of far less political and strategic importance.

able to conclude that rather different motives inform donations in each case. In any event, in 1989–91 around half of all cereal aid donations to the Middle East were received by Egypt; likewise about half of all cereal aid donations to South Asia went to Bangladesh.

However, it may be reasonable to conclude that sub-Saharan Africa is the region where humanitarian considerations most powerfully inform donations of food aid. Only in this world region does cereal aid constitute an appreciable fraction of all cereal inflows (see Table 2.4). Apart from the Middle East, sub-Saharan Africa receives the greatest quantities of cereal aid per head. Also, along with South Asia, the sources of donations to the sub-Saharan region are comparatively widely spread (see Table 2.5). In 1989–91 the largest recipients of cereal aid in sub-Saharan Africa, by far, were Ethiopia and Mozambique. The significance of warfare and the disruption it generates are clear.

Finally in this section, Table 2.4 summarizes the data on cereal production, trade and aid into a single measure of estimated per capita cereal availability.[35] As we would expect, because of traded imports (and to a lesser extent cereal aid) in most regions levels of per capita cereal availability are higher than the corresponding levels of production. This differential between consumption and production is generally quite small, although it is very marked for the Middle East. Conversely, in North America/Oceania levels of per capita availability are far below levels of production. However, Table 2.4 also shows that a large proportion (38 per cent) of world cereal consumption is in the form of livestock feed. Indeed, in North America and Europe a majority of cereal consumption is in the form of feed. In Latin America the figure is around 40 per cent. And in the Middle East cereals roughly equivalent to the region's total traded imports are used as animal feed. Only in sub-Saharan Africa and South Asia are negligible proportions of cereals used to feed livestock.[36]

Because this indirect consumption of cereals, in the form of feed, tends to be much greater in more developed world regions with high overall levels of cereal availability, it follows that variation

[35] Note that here 'availability' covers cereals lost post-harvest, placed in storage, used as seed and animal feed, as well as the quantities eaten by people. In chapter 4 such availability is treated as equivalent to 'consumption'.

[36] Japan and China are the principal users of cereals as animal feed in the Far East, and Brazil and Argentina fulfil similar roles in Latin America. Virtually all of sub-Saharan Africa's use of cereals as feed is accounted for by South Africa and Zimbabwe. South Asia's very low (but increasing) fraction of cereal consumption in the form of feed is partly due to widespread vegetarianism.

between regions in the *direct* consumption of cereals is very much less. While it would be wrong to invest great confidence in the precision of the statistics, this general conclusion is certainly correct. The estimates in Table 2.4 imply that direct cereal consumption varies between about 138 kg per person in sub-Saharan Africa to about 270 kg in the Far East. The fact that the corresponding figures for Europe/FSU and North America/Oceania are only 266 and 234 kg serves to underscore the much greater diversity of diets in these two more developed world regions.

POPULATION GROWTH AND FOOD SUPPLIES

The regional data already examined show that low per capita calorie availabilities tend to be associated with low per capita incomes and high rates of population growth (see Tables 2.2 and 2.3). Clearly, these three variables are interrelated in complex ways. It is not surprising that using our sample countries as units of analysis, variation in per capita calorie availability is more closely associated with variation in income than with variation in rates of population growth.[37] Other things being equal, countries with higher incomes can more readily afford to purchase food. This said, it is too easy to simply attribute 'hunger' as being caused by 'poverty'. For one thing, growth in the food and agricultural sector of the economy has often historically been the main basis of economic growth and higher per capita incomes more generally, i.e. the direction of causation has frequently been partly from improved food output to higher levels of living. For another, while it is true that undernourished populations usually have low per capita incomes, some countries with relatively low incomes have nevertheless been able to achieve comparatively high levels of per capita daily calorie supply.[38]

[37] In a regression of per capita calorie availability on the logarithm of per capita GNP and the rate of population growth for the 91 sample countries 69 per cent of the variation between countries in calorie availability is explained by GNP, whereas the inclusion of population growth, although statistically significant, raises this figure only to 72 per cent. This is because there is a high (negative) correlation between GNP and the rate of population growth, so that although growth on its own accounts for 42 per cent of variation it adds relatively little independent explanatory power.

[38] Countries which have achieved this, by a variety of means, include China, Egypt, Indonesia and Ivory Coast. In many countries raising levels of food output per head essentially constitutes a significant fraction of the task of raising levels of income per head.

For the sample countries Figure 2.2 illustrates the relationship between per capita calorie supplies and recent (i.e. 1980–90) rates of demographic growth. The overall direction of the relationship is obvious. Virtually all national populations growing at less than 1 per cent per year have estimated daily per capita calorie supplies above 2900. It is true that fast-growing populations usually contain higher proportions of young children who generally require fewer calories per head. But this consideration is of only minor relevance to the interpretation of Figure 2.2.[39]

This said, it is important to appreciate that the overall negative association cannot be taken as indicative of a causal relationship in any simple sense. It would probably be as mistaken to conclude from Figure 2.2 that low per capita food supplies were caused by rapid demographic growth, as it would be to conclude that low per capita food supplies cause faster population growth. Instead, what the figure overwhelmingly reflects is differences between countries in the timing of their demographic transitions – and therefore to some extent too, differences in their overall levels of development gauged in the broadest sense. The world's more developed countries (chiefly those of Europe and North America, plus Japan) passed through their periods of rapid population growth (i.e. demographic transitions) many decades ago. In contrast, most developing countries are only now at various stages of their demographic transitions. Virtually all developing nations have experienced dramatic reductions in death rates since the Second World War, but often without commensurate declines in birth rates. Most populations in sub-Saharan Africa and South Asia have probably always experienced very low per capita calorie supplies. Their current low levels of per capita food availability are undeniably associated with rapid rates of demographic growth, but these rates are *not* their principal cause.[40]

The inset in Figure 2.2 shows individual regression lines summarizing the relationships found within each region. In five of the seven cases these relationships are negative – partly reflecting the

[39] It can be shown that the basic relationship remains largely unaltered even if allowances are made for differences in population age-composition.

[40] An allied point worth making is that in many circumstances it makes little sense to say 'poverty causes hunger', because many developing populations have – like the rest of humankind for most of history – always been both poor and undernourished. So instead of focusing on poverty as a 'cause' of hunger it is better to focus on the *processes* which lift people out of poverty, and out of hunger too.

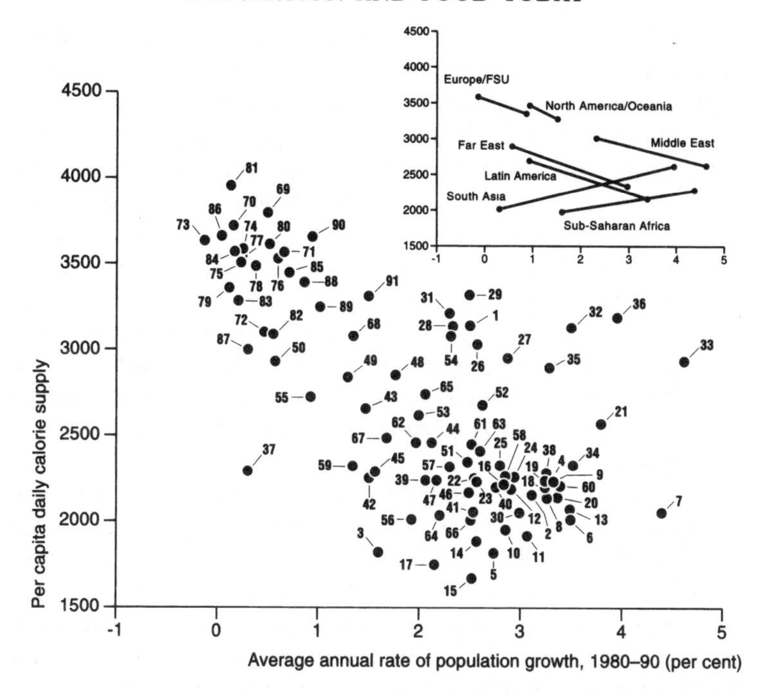

Figure 2.2 Population growth rate and per capita calorie supply around 1990, sample countries

Note: Country numbers are given in Figure 2.1 and in the Appendix

same fundamental point that the least developed countries, with lower per capita food supplies, have tended to come last into the demographic transition. That fast demographic growth does not by itself preclude relatively high levels of per capita calorie supply is suggested by the line for the countries of the Middle East. But perhaps the most interesting feature of Figure 2.2 is the absence of any negative relationship for the countries of South Asia and sub-Saharan Africa. Indeed, in these regions the indicated relationships are marginally positive, i.e. there is a slight hint (no more) that, if anything, faster growing populations tend to have higher per capita calorie availabilities.

The explanation for this distinctiveness of South Asia and sub-Saharan Africa lies partly in the fact that most countries in both regions have extremely low per capita calorie supplies, i.e. there is very little variance with respect to estimated food availability.

Moreover, whereas in some countries in Latin America, the Far East, and even the Middle East, significant birth rate declines have now occurred which have been at least partially informed by rapid socio-economic development, hitherto such development-driven demographic transitions have been comparatively absent from most of South Asia and sub-Saharan Africa. To the extent that falling birth rates have recently lowered population growth rates in some South Asian countries, there is little doubt that government-sponsored family planning programmes have been an important part (though not all) of the explanation. Effective family planning programmes are absent from most of sub-Saharan Africa. And declining birth rates achieved by some countries in the region have hitherto injected little variance into national rates of population growth.

Of course, a host of complex interactions are at work in Figure 2.2. But an examination of the data for the sample countries of either sub-Saharan Africa or South Asia quickly reveals the operation of one important factor which has major implications for both rates of population growth and levels of food supply. This factor is warfare. For example, during the period 1980–90 Mozambique and Afghanistan were respectively the slowest-growing populations in sub-Saharan Africa and South Asia. Due to civil war, large numbers of refugees fled both countries – in the first case in large measure to Malawi, and in the second case overwhelmingly to Iran and Pakistan. These migrant inflows help explain why Malawi, Iran and Pakistan were easily the fastest-growing populations in sub-Saharan Africa and South Asia during 1980–90. Furthermore, the systems of food production and distribution – and hence levels of food consumption – were badly disrupted in the two strife-torn countries, and consequently per capita calorie supplies in each fell to exceptionally low levels even for populations in these regions. And, at least in the case of the refugees from Afghanistan, they fled to countries (especially Iran) with comparatively high levels of per capita food availability.

Alas, Mozambique and Pakistan are not the only examples of these processes, nor are sub-Saharan Africa and South Asia the only regions affected. Even in comparatively developed countries, like parts of the former Soviet Union and Yugoslavia, conflict can have significant simultaneous influences on both population growth and per capita food supplies. And, to the extent that

Figure 2.2 sometimes hints at a positive association between these variables, warfare is certainly germane to the explanation.

We conclude this section with two parting remarks (both of which will be taken up in subsequent chapters). First, while rapid demographic growth cannot be said to have 'caused' the low levels of food availability found in many countries, this does not mean that we can absolve it from the complex of factors which are operating to *keep* these food levels low. Second, there is no doubt that the world's most poorly nourished populations are also generally growing the fastest. It follows that – unless average levels of per capita food supply in these countries can be significantly raised – demographic growth will *by itself* tend to increase the total number of undernourished people and *reduce* the average level of world food supply per head.[41]

AN ASIDE ON FOOD SECURITY

A useful, if rather conveniently imprecise concept for classifying populations according to their vulnerability to hunger and food shortages is that of 'food security'. This can be defined at various levels – for example, with regard to individuals, households or larger populations.[42] But our main concern here is with the country or regional level of analysis. In this context food security can be taken to refer to a population's capacity to obtain adequate quantities of food with a reasonable degree of assurance.

The factors which determine food security depend upon the level of analysis. Household food security can depend upon factors like the age composition of family members, their health status, savings and food stocks, employment prospects, and the prices and agricultural conditions which the household confronts. At the national level food security can ultimately be determined by factors such as the supply of cropland, rainfall variability or a country's capacity to access food through purchases or receipt of food aid. Clearly, these and other determining factors at different levels can combine and interact. For instance, during times of drought the food security of rural households may be conditioned by the operation of national measures of emergency relief (e.g. 'food for

[41] What is being referred to here is sometimes called a 'composition effect'. The idea will recur, for example in chapter 4.

[42] On these issues see Alamgir and Arora (1991), Braun *et al.* (1992), Downing (1992), Pinstrup-Andersen (1994a, 1994b) and World Bank (1986).

work' programmes) which dispense food which has been supplied by other countries.

We cannot address all the factors governing differential national food security here. And no single index can hope to reflect the breadth and complexity of the concept. Nevertheless, using several basic dimensions which have already been introduced, it is worth considering how, in broad terms, the sample countries and regions may differ with respect to their food security status at the start of the 1990s.

Accordingly, four basic indicators have been used to construct a crude food security index (FSI). Other things being equal, a country's level of food security is considered to be high if (i) its level of food availability, measured in terms of its daily per capita calorie supply, is high, (ii) its level of cereal production constitutes a comparatively high proportion of its total cereal consumption, i.e. it is not heavily dependent upon imports (including food aid) to meet its consumption requirements, (iii) it has a high level of per capita income, allowing it if necessary to command food on the international market, and (iv) it is judged to experience a high degree of socio-political stability, thereby minimizing the chance of food shortages resulting from civil disruptions and war. This last dimension of socio-political stability was gauged using an ordinal scale indicator of national political and civil status, ranging from regimes characterized by extreme oppression and pervaded by fear, to open democratic political systems with a free press.[43]

The FSI was computed for each sample country as the average of the normalized values of these four indicators, and broad classes of relative food security status – ranging from 'very high' to 'very low' – were determined by allocating the sample into five groups of

[43] The indicator used is the combined rating of political and civil freedoms produced annually for countries by Freedom House; see Freedom House (1995) and Gastil (1989). The ratings reflect detailed assessments of national institutions and processes (e.g. electoral systems, degree of economic inequality, judicial independence, press freedom, etc.). For reasons of consistency, in calculating the FSI the Freedom House scale was reversed so that countries ranged between 14 (freest) and 2 (least free). Clearly, no single index can capture all relevant aspects of socio-political stability. Also, the implicit assumption is being made here that countries pervaded by tyranny and fear are the most unstable. This may not be true – at least in the short run. However, over the longer run we would maintain that 'open' societies are probably less likely to face major internal disruptions.

equal size.[44] Table 2.6 summarizes the distribution of the sample by relative food security status and region. Country FSI values and relative rankings are given in the Appendix.

The characteristics of this national food security index require some comment. A country's overall level of food security is essentially being gauged as the composite outcome of its level of per capita food availability, degree of cereal self-sufficiency, level of per capita income, and degree of socio-political stability. So only half of the overall index derives from food data *per se*; the other half reflects economic and socio-political conditions which are regarded here as key determinants of overall national food security. Also, definitions of food security commonly embody the idea that food should be available in amounts adequate for a population to be healthy and productive. The present index tries to capture some of this 'absolute' dimension of food security, for example through the estimated level of national per capita calorie availability. However, in practice most populations probably assess the adequacy of their food supplies in relation to the amounts of food which they 'normally' consume. By incorporating measures of the degree of national food self-sufficiency and socio-political stability, the present index attempts to reflect the fact that a country's food security can also be threatened in relative terms – even if its overall level of food availability remains comparatively high. Conversely, a country with low per capita food supplies can still be regarded as having achieved some degree of national food security if its society is stable and self-sufficient in food.

It would be unwise to read too much into the FSI values for individual countries. Nevertheless some interesting general indications emerge from Table 2.6 and the Appendix. As we would expect, all three sample countries in North America/Oceania fall in the 'very high' food security class. A majority of the countries in Europe also fall in this group, although a significant number –

[44] Further information is given in the Appendix. For a similar approach, using a different mix of measures, see Downing (1992). The index of food security for country k (FSI_k) is calculated as:

$$FSI_k = \left(\sum_{j=1}^{4} (I_{j,k} - I_j)/\sigma j\right)/4$$

where $I_{j,k}$ is the value of measure j for country k, I_j is the average value of measure j for all sample countries, and σj is the standard deviation of measure j for all sample countries.

Table 2.6 Distribution of sample countries by relative food security status and world regions

Region	*FSI class*					Total countries	Regional index values
	Very high	*High*	*Intermediate*	*Low*	*Very low*		
Index value =	*>0.68*	*0.68 to −0.14*	*−0.15 to −0.35*	*−0.35 to −0.60*	*<−0.60*		
Sub-Saharan Africa	0	3	5	7	10	25	−0.52
Middle East	0	1	3	2	4	10	−0.34
South Asia	0	0	5	1	1	7	−0.24
Far East	1	2	2	6	0	11	−0.33
Latin America	1	5	4	2	3	15	0.01
Europe/FSU	13	7	0	0	0	20	0.77
North America/ Oceania	3	0	0	0	0	3	1.80
Total	18	18	19	18	18	91	—
Selected countries in class:							
	Australia	Poland	Morocco	Uganda	Kenya		
	Denmark	Portugal	Pakistan	Syria	Algeria		
	UK	Brazil	Iran	Ivory Coast	Burundi		
	Argentina	Turkey	Niger	Sri Lanka	Yemen		
	Japan	Colombia	Zambia	Cameroon	Mozambique		

Notes: (a) The countries listed are for illustrative purposes only. Further details are contained in the Appendix. Given the tentative nature of the FSI values for individual countries, there may be little difference in levels of food security between countries in adjacent FSI classes. Moreover, to reiterate, the level of food security for individual countries may well be influenced by factors which have not been considered here.

(b) Regional indices are population-weighted values of sample countries

including several in eastern Europe plus the former Soviet Union – are classified to be of somewhat lower (though still high) food security status. The countries of the Far East and Latin America are comparatively broadly spread across each of the five classes. Nations in these regions judged to be of 'low' or 'very low' food security are China, Myanmar, Viet Nam, Peru, Cuba and Haiti. It is noteworthy that because of greater socio-political stability and lesser dependence upon cereal imports, the food security levels of countries in South Asia are estimated to be broadly comparable to those in the Far East.[45] Indeed, most South Asian countries are of 'intermediate' food security status. Also of note is that despite their relatively favourable levels of calorie availability the distribution of countries in the Middle East is decidedly skewed towards lower food security – with Algeria, Iraq, Sudan and Yemen all falling in the 'very low' category. Finally, the distribution of sample countries in sub-Saharan Africa is also heavily skewed towards very low levels of food security. Indeed, most of the eighteen countries which emerge as being of 'very low' food security status – including the three with the lowest index values of all (i.e. Mozambique, Somalia and Angola) – are in this region.

While Table 2.6 indicates that all developing regions contain some countries which are subject to low levels of food security, it is also the case that several major developing countries – notably Argentina, South Africa, Brazil, Chile, Turkey, South Korea, Mexico and Thailand – for a variety of reasons fall in the 'high' or 'very high' categories. Turning lastly to the population-weighted regional values, it is unsurprising that North America and Europe both rank high, and that sub-Saharan Africa is the least food-secure region. Latin America does comparatively well. But there is little to choose between the FSI values for the three remaining regions.

CONCLUDING REMARKS

Using a very broad brush this chapter has tried to highlight some of the complexities involved in studying the contemporary 'world

[45] In fact India ranks better in terms of overall food security (41) than does China (63) – although the actual FSI values (respectively −0.23 and −0.45) are not very different. Of course, China's economy is both more prosperous and faster growing than that of India. But it is worth remembering that only as far back as 1959–60 China experienced what was arguably the greatest famine in history. See chapter 3.

food problem'. The empirical basis for such a study is uneven, and sometimes almost nonexistent. Some of the most basic concepts – such as 'food' and 'hunger' – are rather imprecise and difficult to operationalize. Small wonder, then, that depending upon how the data are approached, opinions and disputes can abound.

Yet we have at least identified ways – for example, through aggregation and focusing on cereals – by which some of these difficulties can be addressed. It is obvious that food supplies are especially inadequate in South Asia and sub-Saharan Africa. And for a variety of interconnected reasons the latter region also experiences exceptionally low food security. However, there are some countries which are subject to low levels of both food availability and security in all of the predominantly developing world regions. Poverty is clearly an important influence. But it is far from being the complete answer. Indeed, in many contexts attributing low levels of per capita food availability to poverty may be tantamount to tautology. Hunger and socio-political instability can be at least as closely linked.

The main inequalities of food consumption in the world today rest chiefly in corresponding inequalities of food production. North America/Oceania is obviously the dominant bloc in terms of contemporary food trade and aid. Some populations – notably Egypt and Bangladesh – are dependent to a fair degree upon supplies of food from outside. This said, it is striking how minor is the current aid component of total cereal consumption – even in the case of sub-Saharan Africa. And whatever may be the quantities of food aid given for 'humanitarian' purposes, they are clearly trivial when compared to the quantities used as livestock feed or traded on international markets.

It makes little sense to say that the low levels of per capita food supply found in many countries today have been 'caused' by rapid demographic growth. But the world's worst-fed populations are certainly growing the fastest. Therefore, if the focus is shifted from these average levels of per capita food availability to the *numbers* of people who are poorly supplied with food, then without doubt, rapid population growth is contributing to the total quantum of human hunger.

Conversely, however, it is very doubtful whether the problem of human hunger is having a major independent influence upon contemporary death rates, and thereby rates of population growth. There is no doubt that death rates would be somewhat lower in some poor countries if levels of food availability were improved. But in the modern world a great many factors – lifestyle changes, improvements in medical supplies, immunization programmes, transportation developments, and increased levels of education, to name but a few – combine and interact to reduce mortality. So it is difficult to envisage that changes in food availability alone are having – or in the next few decades will have – a major impact on world population growth. These issues are explored a little more in chapter 3.

3

POPULATION AND FOOD: RECENT TRENDS

INTRODUCTION

This chapter considers trends in population and food since the early 1950s. Particular attention is given to developments during the 1980s and early 1990s, because it is these which lately seem to have generated such alarm in neo-Malthusian circles. Trends during the recent past are also very relevant to our assessment of future prospects which follows in chapters 4, 5 and 6. The initial focus here is on cereals. But subsequently the scope of the discussion will be widened.

TRENDS IN WORLD CEREAL PRODUCTION

Figure 3.1 shows annual world per capita cereal production for the period 1951–93. The historical trend has generally been upward – rising from well under 300 kg per person in the early 1950s to over 350 kg in the period since the mid-1970s. Even at the aggregate world level it is apparent that per capita cereal production is quite variable from year to year. This volatility is due to variation in the global harvest, since the growth of world population can safely be assumed to be comparatively smooth. Note too that there is a hint that the degree of harvest variability may have increased during the recent past.

Figure 3.1 shows that the longer term trend in global per capita cereal production – as gauged by the five-year moving average – has also been fairly changeable. World production of cereals increased significantly faster than the world population during most of the 1950s. But there then followed a period of several years around 1960 when population grew faster than cereal output,

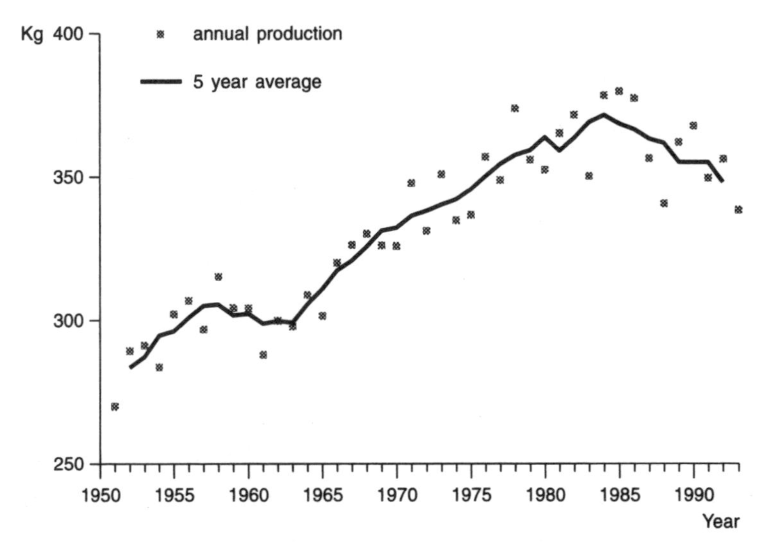

Figure 3.1 World per capita cereal production, 1951–93
Note: Averages for 1952 and 1992 are calculated from data for 1951–53 and 1991–93 respectively

and consequently per capita cereal production declined slightly. Starting from about 1964 there was another period of comparatively brisk gains, perhaps partly marking recovery from the slowdown around 1960. However, the pace of improvement began to slacken from around 1970. And since the early 1980s there has indeed been a fall in world cereal production per head. The moving average curve peaks in 1984 at 371 kg, declining to under 360 kg per person by 1990. So, since about 1984 world population growth *has* been outpacing cereal production (see Figure 3.1).

World per capita cereal production is the outcome of the area of cropland per person which is harvested of cereals and the level of the average cereal yield. Comparing the early 1950s with the early 1990s, Table 3.1 shows that the total global area harvested of cereals rose from about 614 to 696 million hectares. This rather modest 13 per cent increase reflects both the cultivation of entirely new areas of land (for example, in parts of sub-Saharan Africa and Latin America) and increased multiple cropping, i.e. using the same piece of land for more than one crop per year. But because between 1950 and 1990 the world's population more than

Table 3.1 Summary measures of average annual world cereal production, selected periods

	1951–53		*1960–64*		*1970–74*		*1980–84*		*1991–93*
Production (000 mt)	739,695		942,392		1,300,621		1,675,344		1,910,819
Per capita production (kg)	283		300		338		364		349
Total area harvested (000 hectares)	613,719		656,304		689,455		725,145		696,063
Per capita area harvested (hectares)	0.235		0.209		0.179		0.158		0.127
Yield (mt per hectare)	1.205		1.435		1.886		2.311		2.745
Percent of production increase between periods due to: area change		25		13		18		−29	
yield change		75		87		82		129	
		100		100		100		100	
Average annual rate of yield change (%)		1.75		2.73		2.03		1.72	

Note: The allocations of cereal production increase between area and yield changes are approximate. To illustrate the method used, between 1951–53 and 1960–64 average annual total world cereal production increased by 202,697 thousand tons, and the area harvested of cereals increased by 42,585 thousand hectares. Assuming that cereal yields stayed constant at their 1951–53 level (i.e. 1.205 metric tons per hectare) the proportion of total production increase due to area change was about 25 per cent, i.e. ((1.205 × 42,585)/202,697) × 100, leaving about 75 per cent due to increased yields

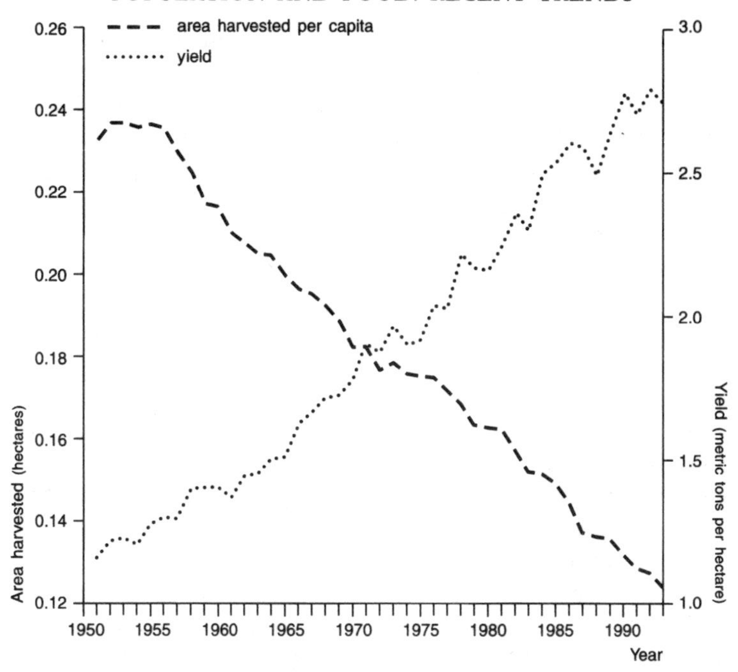

Figure 3.2 World cereal yield and area harvested per capita, 1951–93

doubled, the per capita area harvested of cereals has steadily declined, from about 0.235 to 0.127 hectares per person. On the other hand, during the same period world cereal yields rose from about 1.2 to over 2.7 metric tons per hectare (see Table 3.1).

Figure 3.2 provides eloquent testimony that – quite persistently during the modern era – the average level of world cereal yields has increased, and the per capita area harvested of cereals has declined. Moreover it has to be remarked that Malthus' original characterization that yields increase in an 'arithmetical ratio' (i.e. linearly) is not too far from the mark![1] The modern increase in yields has more

[1] As we noted in chapter 1, when Malthus wrote that 'subsistence increases only in an arithmetical ratio' he was essentially referring to yields, since the discussion was implicitly assuming that the quantity of land was fixed. Note from Figure 3.2 the suggestion that, in fact, measured on a linear scale world yield growth may have accelerated slightly since the mid-1960s; however, with a rising base this has not translated into an increase in the average annual percentage growth rate (see Table 3.1). These issues are examined further in chapter 4.

than offset the decline in per capita area harvested, thus producing the general rise in world per capita cereal production. In fact, about 90 per cent of the increase in the size of the global cereal harvest since the early 1950s has been due to increases in yield, and only about 10 per cent has been due to increases in total harvested cereal area. Furthermore, since the early 1980s changes in yield have been entirely responsible for rises in the global harvest. This is because the total area of cropland harvested of cereals has declined quite sharply (see Table 3.1).

To examine trends in per capita cereal production, harvested area and yield a little more closely, some further degree of magnification and integration is required. Accordingly, Figure 3.3 shows trends in the corresponding average annual percentage growth rates, based upon two consecutive moving averages of annual observations, each of five years in length.[2] These growth rates reflect variation in global production, area and yield change over the short-to-medium run. Note too that the average annual growth rate of per capita cereal production (i.e. the bold line) is equal to the sum of the other two growth rates. So it is easy to see how per capita production has been influenced by changes in per capita area harvested and yield.

As we would expect, throughout the period under review the percentage growth rate of per capita area harvested of cereals has been negative, and the yield growth rate has been positive (see Figure 3.3). Except for the brief interval around 1960, and the period since 1984, the overall sum of these two growth rates has been positive – and therefore so too has been the growth rate of world per capita cereal production. The rates of change of both area and yield have varied over time. But it is clear that alterations in the growth rate of per capita cereal output have been mainly determined by changes in the growth rate of yields – confirming

[2] To be precise, moving averages of per capita area harvested and yield were calculated exactly as has been done for per capita cereal production in Figure 3.1. Then for each of these three smoothed series, average annual growth rates were calculated for consecutive periods of five years in length, with one year overlap. Because middle years were repeated in this procedure, the resulting annual growth rates are each based on nine years of observation and are plotted on the fifth (i.e. central) year. The growth rates shown for 1954 and 1990 are based on seven years of observation. It should be clear that use of moving averages diffuses period effects (e.g. of a crisis) over several years, and that periods of relatively low growth rates are likely to be followed by periods of relatively high growth rates (reflecting recovery) and vice versa.

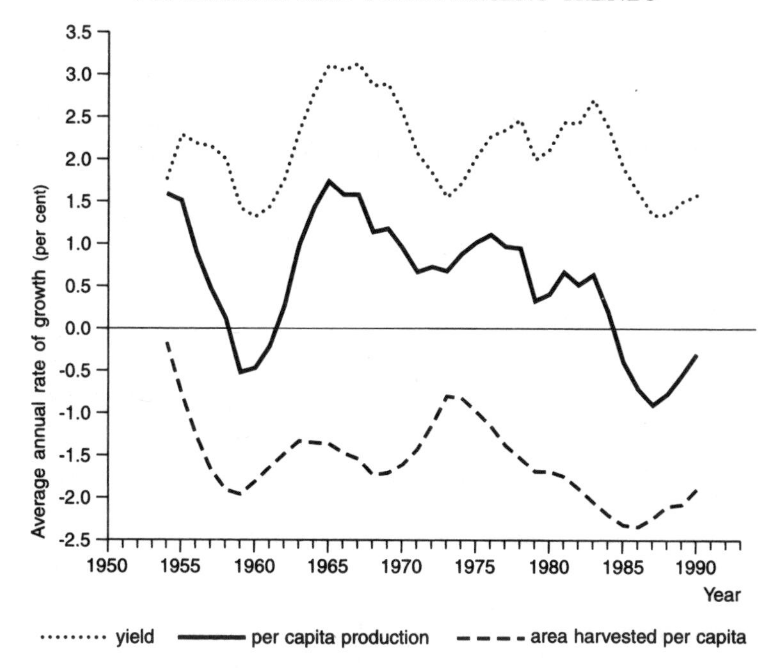

Figure 3.3 Average annual growth rates of world per capita cereal production, area harvested and yield, based on five year averages

the previous suggestion as to their primary importance. The decline in per capita cereal production since the mid-1980s clearly partly reflects some recent moderation in yield growth rates – although these growth rates have generally remained just within the previous range of variation. But what is really distinctive about the decline in per capita cereal output of recent years is the contribution made by an exceptionally negative rate of change of harvested area during the 1980s (see Figure 3.3). Indeed, we have already seen that the total world area harvested of cereals has recently been dramatically reduced (Table 3.1).

The period of the 1980s and early 1990s will be considered later. First, however, some observations are required regarding the longer term trends since the early 1950s.

WHERE IS THE 'GREEN REVOLUTION'?

A 'revolution' usually signifies a sudden and dramatic change. The term 'green revolution' is generally used to refer to the important developments in food production which stemmed from the release of new high yielding varieties (HYVs) of inbred rice and wheat in the late 1960s. The rice varieties owed much to research carried out at the International Rice Research Institute (IRRI) in the Philippines, and the wheat (and later maize) varieties owed much to work at the Centro Internacional de Mejoramiento de Maiz y Trigo (CIMMYT) in Mexico.

The crucial feature of these new HYVs was their higher yields compared to traditional cereal varieties. These improved yields were mainly achieved because the grain head constituted a larger proportion of the total plant – the stems being shorter, but sturdy.[3] In addition, some new varieties are less sensitive to the length of daylight – thus increasing the geographical range of locations over which they can be grown. And HYVs generally mature faster than traditional cereals – so increasing the scope for multiple cropping. To achieve their potential, most HYVs require substantial inputs of chemical – especially nitrogen – fertilizers, plus irrigation water which greatly facilitates fertilizer up-take by the plant.

As with most technical and scientific developments, the passage of time has revealed various problems and constraints. For example, because HYVs are monocultures an entire crop can be threatened by pests, so increasing the need for (costly) pesticides. With their shorter stems, in some locations HYVs have been especially prone to destruction by flooding. Other new varieties have been particularly susceptible to disease. It has even been suggested that these developments may partly explain the seeming recent increase in variability of world cereal production, since in the past greater crop diversity may have helped produce greater overall harvest stability.[4] More generally, the increased fertilizer and irrigation requirements of HYVs have sometimes limited their adoption by poor farmers.

However, in view of their decisive advantage of higher yields, the spread of new varieties was often very fast, especially in the early stages of introduction, and particularly in locations in Asia and

[3] Early HYVs were generally stiff-strawed, and of 'dwarf' or 'semi-dwarf' stature.
[4] See Pinstrup-Andersen (1994a, p. 6). It is clear from pp. 87–9 below that an increasing frequency of severe drought is the chief explanation for increased harvest variability.

Latin America. In 1964–65 there were only about 80 hectares of cropland planted with HYVs in the whole of Asia. By 1968–69 this figure had increased to 13.7 million hectares. Since the early 1990s almost 100 million hectares of land in Asia have been planted each year with modern varieties of rice alone.[5]

These 'green revolution' advances have certainly had an important effect on yields, although their precise contribution is extremely difficult to gauge. However, a striking feature of Figures 3.1, 3.2 and 3.3 is the *absence* of any obvious sign of a revolution in the sudden transforming sense of the term. Increases in per capita cereal production and yields have generally been similar in the periods before and after the late 1960s.[6] This statement holds for most world regions. Indeed, examination of the data for the 91 sample countries reveals very few cases where national per capita cereal output increased abruptly onto an entirely new plane solely due to the sudden introduction of HYVs. Because a very high proportion of its cereal production is of wheat, and because an exceptionally large fraction of its cropland is irrigated, the introduction in the late 1960s of new Mexican wheat varieties into Pakistan probably constitutes the best – and certainly a very rare – example of such a national revolution (see Figure 3.4).

Hence the question: 'Where is the green revolution?' The answer, in brief, is that it is 'everywhere and nowhere' in Figures 3.1, 3.2 and 3.3. The late 1960s was a period of significant developments. But better yielding wheat varieties were being introduced in North America and Europe in the 1950s, and even earlier. And much the same was true for rice in parts of Asia, notably Taiwan and Japan. In recent decades agricultural progress has been a continuous, multifaceted and increasingly international process. Some analysts have suggested thinking in terms of two or three overlapping 'green revolutions'.[7] But even this may be misleading. In many countries several cereal crops are grown, and 'revolutionary' changes with respect to a particular crop may be obscured in national data. Even where one type of cereal crop (e.g.

[5] For these figures see IRRI (1993, p. 130) and Kellman (1987, p. 18).

[6] The relatively high yield growth rates shown in Table 3.1 for the period between 1960–64 and 1970–74, and in Figure 3.3 for the mid-1960s, are both mainly explained by 'catch-up' from China's food crisis in the years around 1960 (see pp. 68–9).

[7] See Plucknett (1993, pp. 13–14). The first HYV cereal release (Norin 10) occurred in 1935 and originated from earlier Japanese research. See Smil (1991, p. 598).

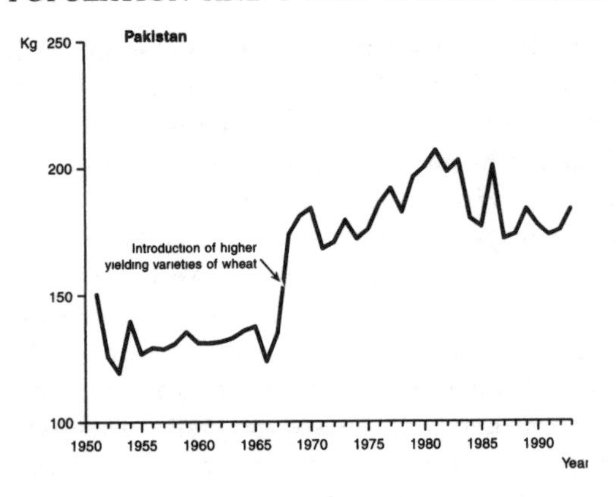

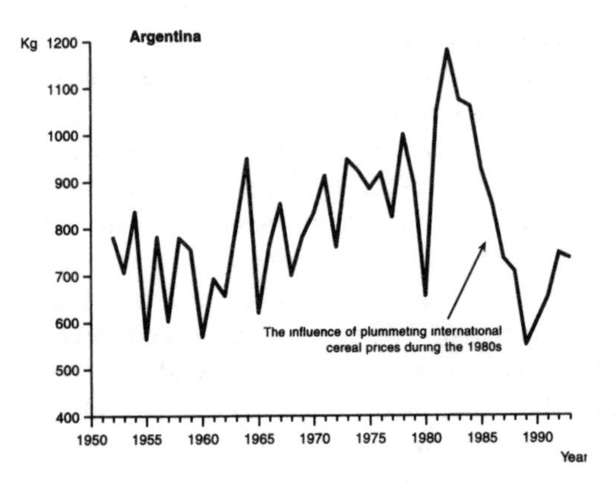

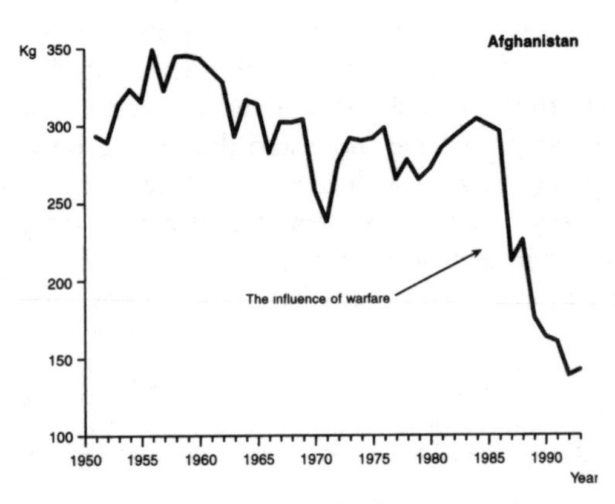

Figure 3.4 Per capita cereal production in selected sample countries, 1951–93

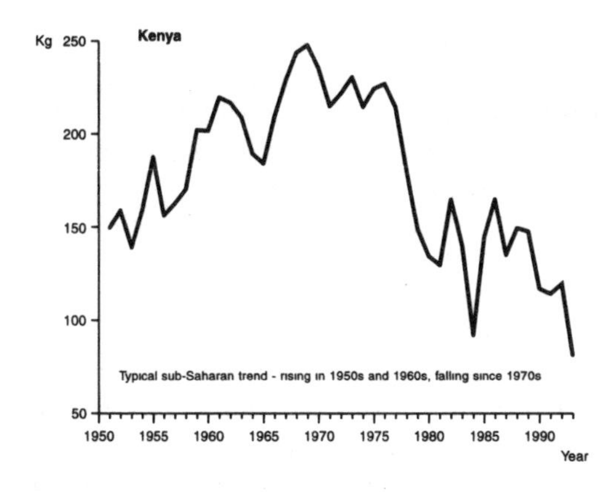
Kenya
Kg
250
200
150
100
50
Typical sub-Saharan trend - rising in 1950s and 1960s, falling since 1970s
1950
1955
1960
1965
1970
1975
1980
1985
1990
Year

Saudi Arabia
Kg
300
250
200
150
100
50
0
Massive investment in
irrigated wheat production
1950
1955
1960
1965
1970
1975
1980
1985
1990
Year

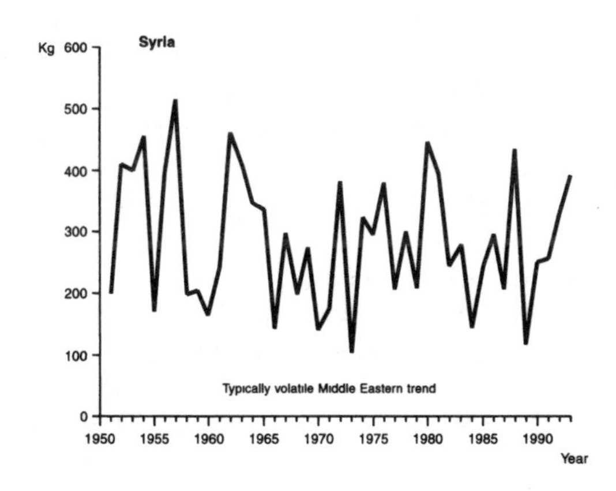
Syria
Kg
600
500
400
300
200
100
0
Typically volatile Middle Eastern trend
1950
1955
1960
1965
1970
1975
1980
1985
1990
Year

rice) predominates, it may be grown under a variety of agro-climatic conditions, each requiring different HYVs and moderating the speed of their spread. Furthermore, the term 'green revolution' arguably distracts attention from a host of important collateral developments – such as better crop protection, improved understanding of plant nutrients, the spread of tubewells and rural electrification – all of which contribute to yield growth and the overall process of agricultural advance.

In summary, the term 'green revolution' has probably contributed to a misrepresentation and oversimplification of global food production trends during recent decades. These trends are better conceptualized as the outcome of a complex and continual *process* involving investments of effort and resources, by many different actors, in research, innovation, education and on the farm.

FOOD CRISES AND FAMINES

Most recent food crises and famines have been comparatively localized events. So it may seem a little odd to discuss them in the context of global trends. However, the statistical time series in Figure 3.1 and, still more, Figure 3.3 do show the marks of such crises. Therefore the subject deserves some attention here.

The disaster in China in 1959–64

The main cause of the 'plateau' in world cereal production around 1960 and the concurrent decline in the growth rate of global cereal yields was a calamity in China (see Figures 3.1 and 3.3). Following a fair harvest in 1958, China's then modest level of per capita cereal production plummeted by 30 per cent to reach a minimum in 1961. Several years of slow recovery followed. But it was not until the early 1970s that per capita cereal output in China revived to pre-crisis levels.

The causes of this production failure were complex and lie largely beyond the present work. Prolonged drought and heavy floods during 1959–61 were certainly significant contributory factors. But overshadowing them in importance were the massive social, political, economic and agricultural disruptions resulting from the Chinese Communist Party's so-called 'Great Leap Forward'. Estimates of the total number of excess deaths during the crisis vary between 16 million and 30 million; a plausible intermediate number is 23 million.[8] There was also a large reduction in the number of live births.

Before the disaster China accounted for about one-quarter of both the world's population and cereal production. So as well as leaving a clear imprint on food production trends (e.g. see Figure 3.1), the crisis also affected the volume of global demographic growth. The United States Bureau of the Census estimates that, comparing 1956–58 and 1959–61, the average annual increment of the world's population fell from about 55 million to 46 million.[9] Although this disaster was restricted to China – and indeed remained largely hidden from the rest of humanity at the time – its sheer scale alone might well justify use of the epithet 'a global crisis'.

The 'world food crisis' of 1972–74

The world food crisis of 1972–74 is actually surprisingly hard to discern in the time series of global per capita cereal production. In 1972 world cereal output per head fell by some 5 per cent from the previous record level – set in 1971. Then in both 1974 and 1975 per capita output was down by a similar margin compared to the new record level – set in 1973. But there is little in Figure 3.1 to indicate why there was a so-called 'world crisis' in the early 1970s compared to other periods.

This crisis was caused by a combination of circumstances. By 1969–70 there was some confidence about future world food prospects – partly due to the then highly publicized benefits seemingly promised by the 'green revolution'. World cereal stocks – expensive to maintain and disproportionately held in North America – were at comparatively reasonable levels. Therefore the United States and Canada both encouraged their farmers to 'idle' or 'set aside' large areas of cereal cropland.

Then in 1972 several major world grain producing regions were simultaneously hit by drought – an occurrence which has been linked to the oceanic-atmospheric circulation in the Pacific Ocean which is now known as the El Niño–Southern Oscillation (ENSO). These droughts affected some heavily populated agricultural areas,

[8] See Peng (1987). In this context excess deaths are those which would probably not have occurred in the absence of a crisis.
[9] For these estimates of annual additions to the world's population see Starke (1993, p. 95).

like northern and western parts of South Asia and parts of north-eastern China. Faced with a significant production shortfall, the former Soviet Union secretly purchased large amounts of wheat on the international market, and cereal prices began to rise. In turn, countries in the Middle East, already sizeable purchasers of cereal imports, were spurred to raise the price of oil by limiting its production.[10] World per capita cereal production reached a new record in 1973 – as major producers like the United States, Canada and the Soviet Union expanded their planted cereal area in order to help cope with the unfolding situation. However, 1974 also saw a drop in the global cereal harvest (see Figure 3.1).

But if the time series on world per capita cereal production conveys only a weak impression of a global crisis, many individual countries certainly experienced an abrupt change in their cereal production. For the 67 developing countries in our sample Figure 3.5 shows the proportion each year which experienced levels of per capita cereal output above their long-run linear trend.[11] Whereas in 1970 nearly two-thirds enjoyed 'above-trend' production, in each of the years 1972, 1973 and 1974 the fraction was well under a half.

The crisis is even more discernible in Figure 3.6. Because of increased purchases, and speculation, international cereal prices rose sharply in 1973 and had more than doubled by 1974. High prices in turn stimulated a revival in world per capita cereal area (see Figure 3.3); indeed, the total area of cropland harvested of cereals in 1975 was 6 per cent greater than in 1970. It was not until 1977 that world grain prices returned to their previous levels. The crisis also left its mark on world cereal stocks, which fell from over 20 per cent of annual consumption in 1970 to around 15 per cent

[10] On the distribution of droughts in 1972 (and 1973) and the possible links to ENSO, see Glantz (1991). For a summary of other aspects of the crisis – such as the creation of FAO's Global Information and Early Warning System (GIEWS) and the 1974 World Food Conference and World Population Conference, see for example Kutzner (1991, pp. 84–7). Analysts such as Mitchell and Ingco consider the oil price rise – which had the knock-on effects of raising fertilizer prices, and increasing the demand for food of oil-exporting countries like Indonesia and Mexico – to be the most important single cause of the crisis; they contend that the combination of causal factors in 1972–74 was largely coincidental. See Mitchell and Ingco (1993, pp. 3–4). Perhaps. It is true that the world oil price would probably have risen at some time in any event. But it is a moot point whether the timing, speed, extent or consequences of the rise would have been as serious had the droughts of 1972 and 1973 not occurred.

[11] For each country the reference trend used as the basis for Figure 3.5 was a straight line fitted to per capita cereal output for years 1951–93 inclusive. In 34 of the 67 cases the reference trend was itself negative.

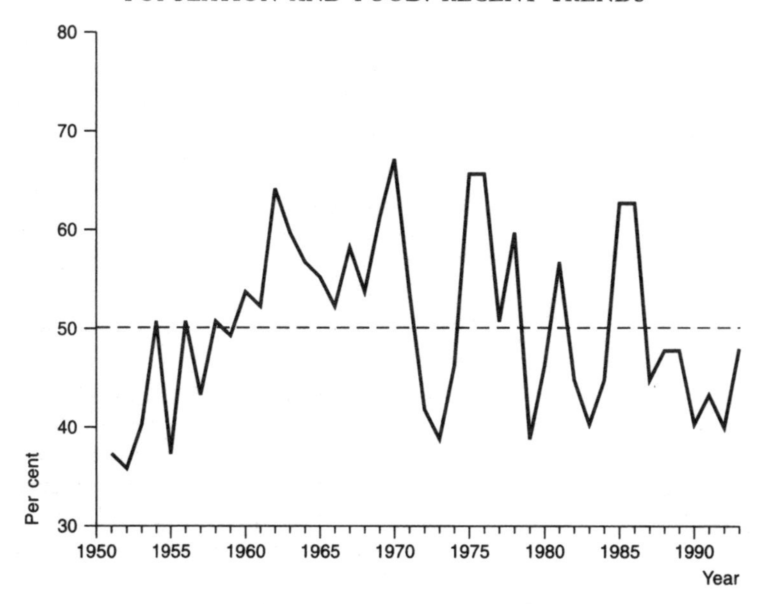

Figure 3.5 Per cent of developing countries with per capita cereal production above long-run trend, 1951–93

in 1973, 1974 and 1975. Lower stocks meant sharply reduced volumes of cereal aid donations. The basic relationships linking movements in world cereal prices, stocks and donations are very clear from the moving averages in Figure 3.6.

It is hard to be precise about the demographic consequences of this crisis. But they were certainly very much smaller than those of the 1959–64 disaster in China. In 1972–74 the worst affected poor populations were those of Ethiopia, the Sahel (both regions for which demographic data are largely lacking), and parts of the Indian subcontinent and China. A very rough estimate of excess mortality in Ethiopia in the entire period 1968–75 is half a million.[12] The most informed and widely cited estimate for the Sahel is that no more than a quarter of a million excess deaths occurred during the period 1970–74.[13] Excess deaths in India in 1972 and 1973 may have been between 500,000 and 1 million, though the

[12] For this figure see Braun (1991, p. 3).
[13] This is an educated guess. See Caldwell (1975, p. 48). Prominent Sahelian countries in our sample are Burkina Faso, Chad, Mali, Niger and Sudan.

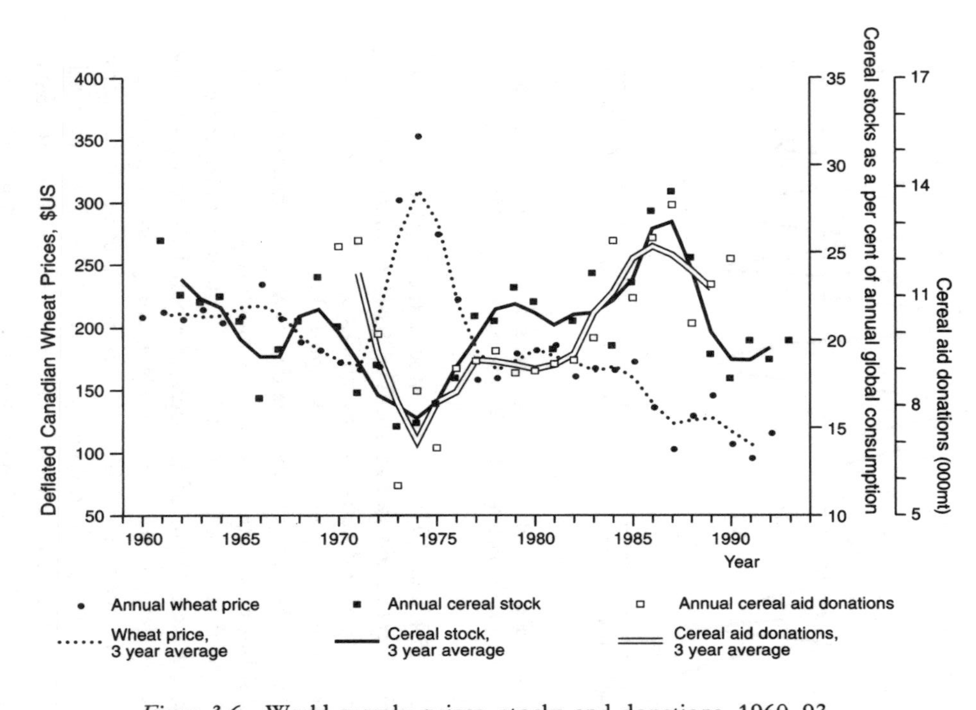

Figure 3.6 World cereals: prices, stocks and donations, 1960–93

Principal sources: Canadian wheat prices are from Mitchell and Ingco (1993, p. 179). Stock data are from USDA, see Brown *et al.* (1994). Data on cereal donations are from FAO (various years) *Food Aid in Figures*, Rome.

basis for this statement is not strong.[14] In addition, there were probably several hundred thousand excess deaths in Bangladesh during 1974 and 1975, precipitated by flooding rather than by drought.[15] The available death rates for China suggest that in 1972 there may have been 200,000 or 300,000 excess deaths.[16] In all these locations there was probably also a significant reduction in the number of live births. But, comparing 1969–71 and 1972–75, the US Bureau of the Census estimates suggest that the annual global increment in population fell only from about 77 million to 75 million people per year.[17]

The crisis of 1983–84

The events of 1983–84 carry echoes of the crisis of 1972–74. In the early 1980s world cereal stocks were again deemed to be rather high. To limit output, in 1983 the United States withdrew a massive 31.5 million hectares of cropland from production.[18] But 1983 also proved to be a drought year in the United States. And both 1983 and 1984 were years of severe drought in much of sub-Saharan Africa and parts of the Middle East; again, the ENSO anomaly has been implicated. World per capita cereal production fell quite sharply in 1983, and global cereal stocks fell quite sharply in 1984 (see Figures 3.1 and 3.6). Virtually none of the 25 sample

[14] According to one set of unadjusted figures (see Bhat *et al.* 1984, p. 70) the death rate in India in 1972 was 1.3 points above the average figure for 1970 and 1971; this might very roughly imply an extra 600,000 deaths. Another estimate based on somewhat better data is that in the Indian state of Maharashtra there were about 130,000 excess deaths in 1972–73. See Dyson and Maharatna (1992, p. 1328). The total national figure is unlikely to have been much greater than 1 million.

[15] One widely quoted estimate by Alamgir (1980, pp. 142–3) is excess mortality of 1.5 million. But this figure is based on survey data relating to only a few hundred deaths and therefore is highly contestable. Data for the Matlab area of Bangladesh suggest that the death rate there rose from about 16.4 per thousand in 1973–74 to 20.8 in 1975, which if applied nationally would indicate around one-third of a million excess deaths. See Dyson (1991, p. 286).

[16] Some more crude calculations: the official death rate in China in 1972 was 7.61 compared to 7.32 the previous year. See Coale (1984, p. 69); see also Banister (1987, p. 228). In a national population of 880 million, this would indicate around a quarter of a million extra deaths.

[17] See Starke (1993).

[18] See USDA (1992, p. 8). To emphasize the importance of the US in world agriculture, note that 31.5 million hectares is equivalent to either the total land surface of Poland, or 38 per cent of the total global increase in area harvested of cereals between 1951–53 and 1991–93 (see Table 3.1).

countries in sub-Saharan Africa experienced 'above-trend' per capita cereal output during the years 1982–84, an effect which is discernible in the data for the total developing country sample (see Figure 3.4).

However, because Asia was largely unaffected, this was a much smaller crisis even than that of 1972–74. World cereal prices barely rose at all (see Figure 3.6).[19] There certainly was some excess mortality in sub-Saharan Africa, especially in Ethiopia and Sudan where civil strife must have contributed hugely to the death toll.[20] But it is extremely difficult to see that the 1983–84 crisis had any sizeable impact on the volume of world demographic growth.

Postscript on famines

We conclude this section with three observations. First, a decline in food availability – usually because of harvest failure – *is* a significant contributory causal factor to most modern famines.[21] Of course, many other factors can also be involved, ranging in Adam Smith's words from 'the violence of government' to the 'waste of war'. But 'real scarcity' due to the 'fault of the seasons' is usually a part of the causal jig-saw puzzle too.[22] This was true in

[19] There were slight price increases in 1983 and 1984 for maize and sorghum, which are both 'preferred' sub-Saharan cereal grains. See World Resources Institute (1992, p. 242).

[20] See Braun (1991), Teklu *et al.* (1991) and Webb *et al.* (1992).

[21] There is a misleading impression in much of the literature that a sizeable proportion of modern famines do not involve a food availability decline (FAD). This belief has arisen partly due to Amartya Sen's influential analysis of five famines in his book *Poverty and Famines* (Sen, 1981). In two cases – the Sahel famines of the early 1970s and the 1974 famine in Ethiopia – Sen acknowledges that FADs occurred. But in the remaining three cases – the famines in Bengal in 1943–44, Bangladesh in 1974–75 and Wollo (Ethiopia) in 1972–73 – Sen contends that there was no FAD. See Sen (1981, p. 154; 1990, p. 37). However, several subsequent detailed studies have cast considerable doubt on Sen's conclusions. Goswami's very careful reanalysis for Bengal shows convincingly that 'food availability in 1943 was less than in 1941, contradicting Sen's finding that FAD could not have been important.' See Goswami (1990, p. 457). Basu's reanalysis of the Bangladesh crisis concludes that 'the argument of Sen that the total availability of food in the year of the famine was not far from normal cannot be true.' See Basu (1984, p. 295). Finally, Kumar writes of Wollo that 'Insofar as a binding transport limitation accentuated the chronic food shortages caused by drought in the province, then, food availability decline has to figure as the major explanatory factor in the famine.' See Kumar (1990, p. 184). In conclusion, while they are far from being complete explanations, FADS were probably involved in all five of Sen's famines.

[22] For these quotations see Smith (1776, p. 109).

1959–64, in 1972–74 and in 1983–84. Today, probably only warfare is a more common and easily identifiable immediate cause of famine.

Second, even from the foregoing overview, it is clear that the frequency and demographic impact of famines has declined greatly during recent decades. Modern food crises have had a diminishing and increasingly short-lived impact on population growth. The phenomenon of famine has become more and more restricted to sub-Saharan Africa. And even there, the chances of someone dying in a famine have been described as 'vanishingly small'.[23]

Finally, of course, this in no way reduces the need to avoid such dreadful events. The element of 'replay' in 1983–84, compared to 1972–74, underscores the need for vigilance.

REGIONAL CEREAL PRODUCTION TRENDS

Explaining the decline in world per capita cereal production since 1984 – and the relatively slow yield growth rate and contraction of harvested area which we have seen are its proximate causes – requires a regional perspective. An appreciation of movements in international cereal prices and stocks during the 1980s is also needed. Regional trends in per capita cereal production are shown in Figure 3.7; and Figure 3.8 decomposes the regional growth rates of per capita cereal output according to the rates of change of per capita area harvested and yield.[24] Relevant regional statistics are given in Table 3.2. For convenience and brevity the world regions will be considered in three groups.

Sub-Saharan Africa and the Middle East

The long-term trend in sub-Saharan Africa's per capita cereal production arguably reflects much of the region's modern history (see Figure 3.7). The 1950s were comparatively settled times administratively, rainfall was quite good and cereal output per person increased. But starting in the 1970s, after most countries had gained their political independence, a fairly steady decline set in. This decline seems to have quickened since about 1980, partly

[23] See Seaman (1993, p. 31). He stresses the paucity of evidence, and the problem of specifically distinguishing famine mortality in the region. See also de Waal (1993).

[24] Figures 3.7 and 3.8 are comparable to Figures 3.1 and 3.3.

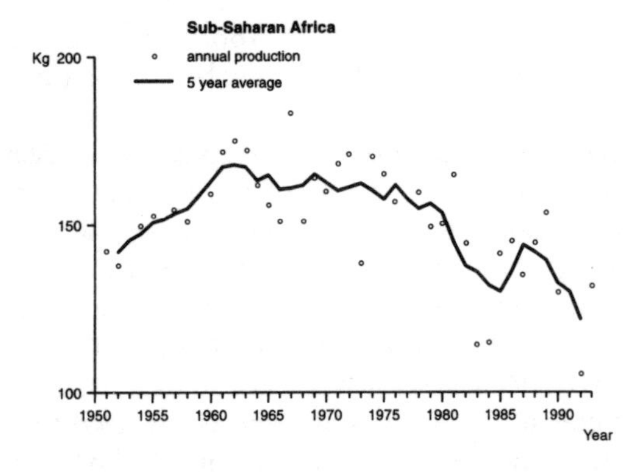

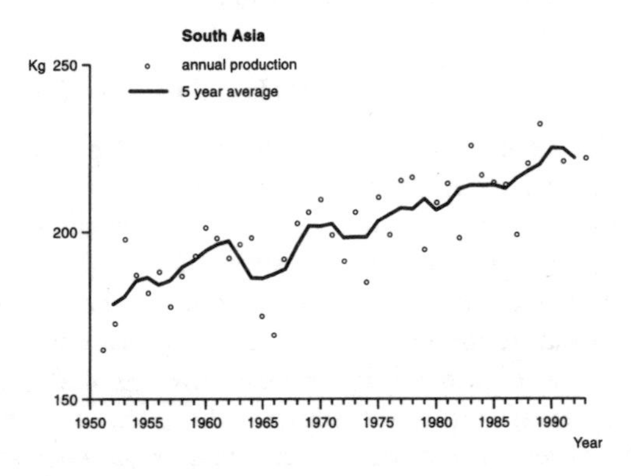

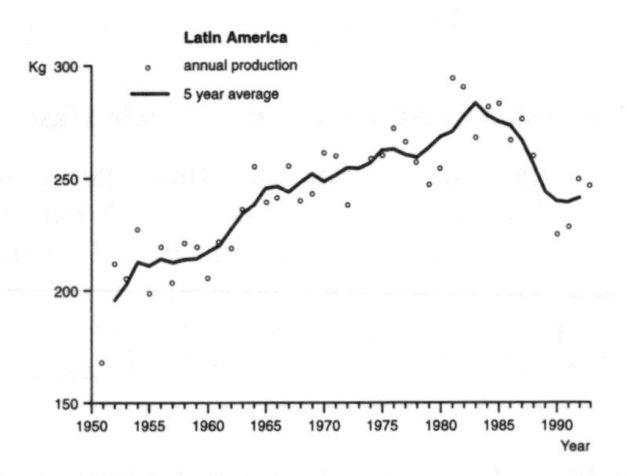

Figure 3.7 Per capita cereal production by world region, 1951–93
Note: Averages for 1952 and 1992 are calculated from data for 1951–53 and 1991–93 respectively

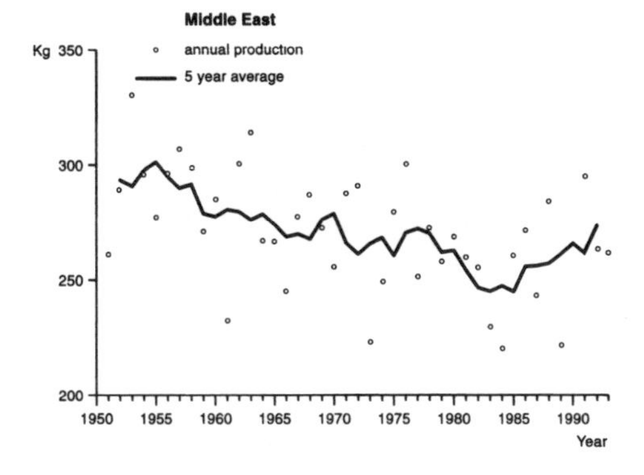
Middle East
Kg 350
annual production
5 year average
300
250
200
1950 1955 1960 1965 1970 1975 1980 1985 1990
Year

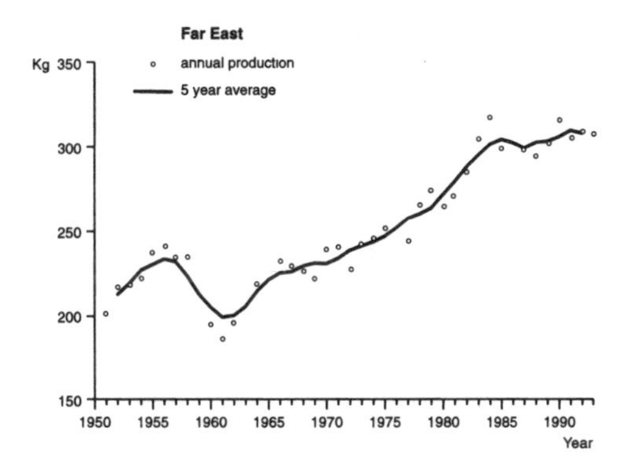
Far East
Kg 350
annual production
5 year average
300
250
200
150
1950 1955 1960 1965 1970 1975 1980 1985 1990
Year

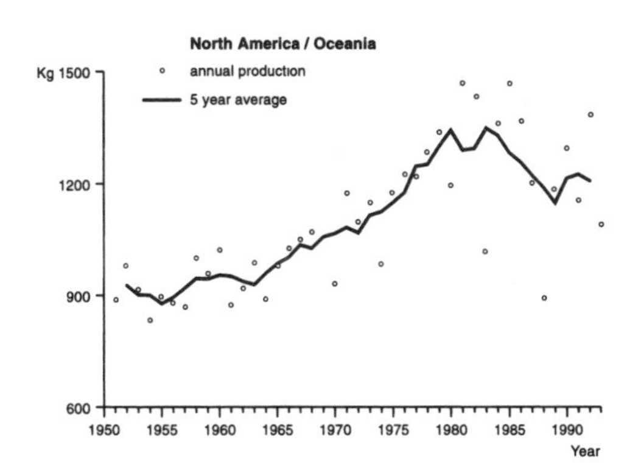
North America / Oceania
Kg 1500
annual production
5 year average
1200
900
600
1950 1955 1960 1965 1970 1975 1980 1985 1990
Year

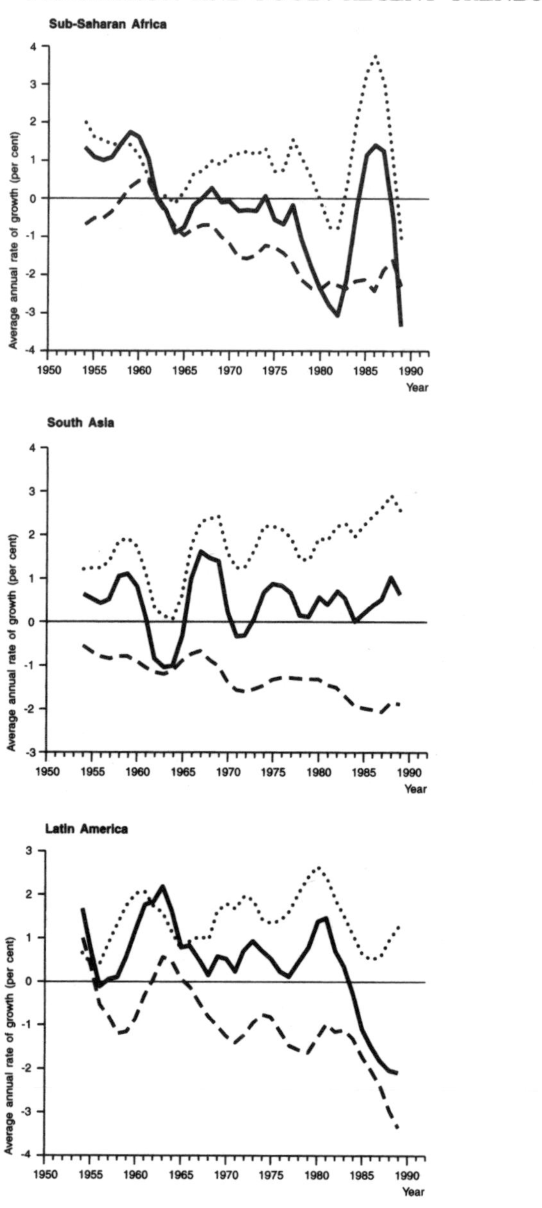

Figure 3.8 Average annual growth rates of regional per capita cereal production, area harvested and yield, based on five year averages

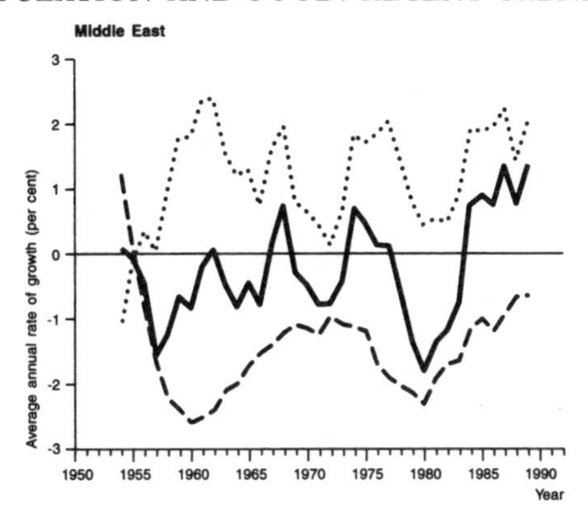
Middle East
Average annual rate of growth (per cent)
3
2
1
0
-1
-2
-3
1950 1955 1960 1965 1970 1975 1980 1985 1990
Year

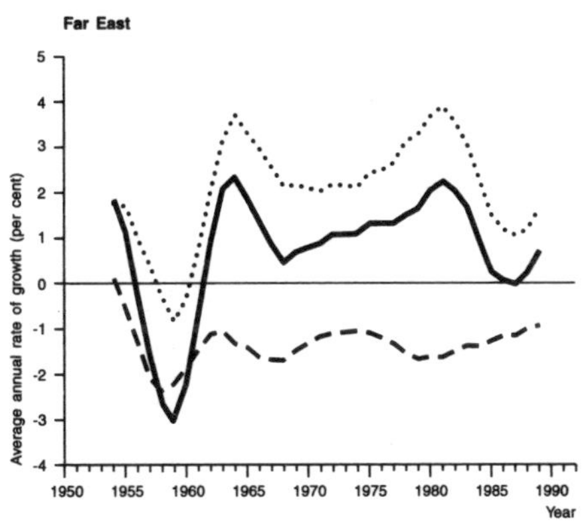
Far East
Average annual rate of growth (per cent)
5
4
3
2
1
0
-1
-2
-3
-4
1950 1955 1960 1965 1970 1975 1980 1985 1990
Year

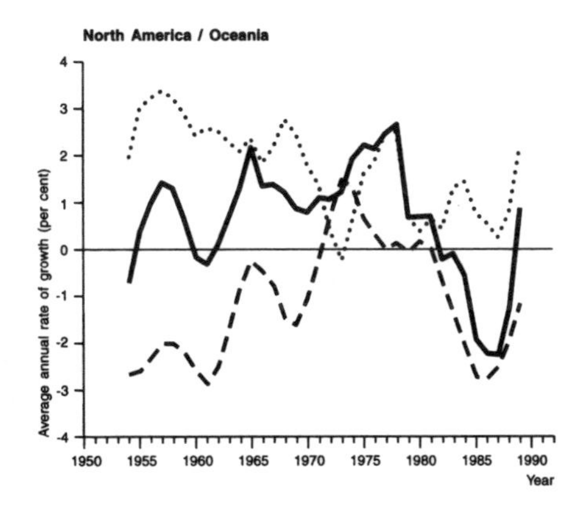
North America / Oceania
Average annual rate of growth (per cent)
4
3
2
1
0
-1
-2
-3
-4
1950 1955 1960 1965 1970 1975 1980 1985 1990
Year

due to the severe droughts of 1983, 1984 and 1992. Kenya is a fairly typical example (see Figure 3.4). Cereal yield growth rates in the sub-Saharan region have generally been positive. But the brief spurts in the percentage growth rates of both per capita production and yields in the mid-1980s merely reflect short-term recovery from the crisis of 1983–84 (see Figure 3.8). Despite a 3 per cent increase in total area harvested of cereals between 1980–86 and 1987–92, the area harvested per person is rapidly declining given the region's exceptionally fast rate of demographic growth (see Table 3.2).

These adverse trends in sub-Saharan Africa are widespread and have many dimensions – including overvalued exchange rates, poor transport, marketing, storage and support services, common and longstanding neglect of the agricultural sector by government, and the importation of cheap supplies of wheat from the international market in order to feed potentially volatile and rapidly growing urban populations. Overall, and given present circumstances, the evidence is fairly convincing that this region is incapable of expanding its cereal production to match its population growth.

Over the long run, per capita cereal output has also declined in the Middle East (see Figure 3.7). Yield growth rates have been positive, but insufficient to offset the decline in per capita harvested area. In this region too the early 1980s were especially difficult because of droughts. But since 1985 the trend in per capita cereal production has been a little more promising (partly due to events in Saudi Arabia – see Figure 3.4). The recovery in the yield growth rate in the Middle East appears to be relatively sustained, rather than simply a rebound from crisis (see Figure 3.8). The region's total area harvested of cereals rose by some 9 per cent between 1980–86 and 1987–92 (see Table 3.2). So the recent improvement in per capita cereal production reflects both better performance in yields and expansion of area.

South Asia and the Far East

The time series for South Asia and the Far East show clear signs of previous major food crises – especially those in India in 1965–66 (and to a lesser extent 1972–74) and, of course, China in 1959–64 (see Figure 3.7). These early crises were marked by large reductions in the growth rates of per capita production and yields, and were followed by strong upswings associated with recovery (see Figure

Table 3.2 Regional measures of recent changes in cereal production

Region	*Average area harvested (million hectares)*		*Change in area (%)*	*Average annual rate of change (%) of cereal yield 1951–57 to 1980–86*	*Average annual rates of change (%) 1980–86 to 1987–92*			*Change in yield growth rate after 1980–86*
	1980–86	*1987–92*			*Per capita production*	*Area harvested per capita*	*Yield*	
Sub-Saharan Africa	53.1	54.7	+3.0	1.08	−0.70	−2.60	1.87	+0.79
Middle East	35.8	39.1	+9.2	1.41	0.59	−1.42	1.99	+0.58
South Asia	141.8	139.4	−1.7	1.98	0.48	−2.51	2.99	+1.51
Far East	142.0	142.6	+0.5	2.68	0.65	−1.47	2.09	−0.59
Latin America	51.8	49.8	−3.9	1.91	−1.41	−2.61	1.23	−0.68
Europe/FSU	185.3	172.4	−6.9	2.51	0.16	−1.63	1.74	−0.77
North America/ Oceania	109.3	96.7	−11.5	2.41	−1.77	−2.90	1.13	−1.28
World	724.1	700.2	−3.3	2.25	−0.52	−2.26	1.72	−0.53

Notes: (a) The data on regional harvested area relate to gross harvested area in the sample countries only; all such data, especially for regions like sub-Saharan Africa, should be regarded with caution. Regional yield *levels* are addressed in chapter 4

(b) For the period 1980–86 to 1987–92, note that the average annual rates of production change are approximately equal to the sum of the rates of change of per capita harvested area and yield (discrepancies being mostly due to interaction and rounding effects)

3.8). However, in both regions the general trend in per capita output has been rising, as yield increases have more than offset declines in per capita area harvested. Yield growth rates over the long term have been particularly impressive in the Far East – largely due to the performance of China, starting from the 1970s. During the 1980s per capita cereal production continued to rise in both regions. It is true that average annual percentage yield growth rates in the Far East have recently declined quite sharply. But this is partly a consequence of the strong spurt in yields attained in China in the early 1980s, following major agricultural policy reforms in 1978.[25] In South Asia recent yield growth rates have attained unprecedented levels (see Figure 3.8). In both these world regions the total area harvested of cereals remained roughly constant between 1980–86 and 1987–92. So the faster population growth of South Asia accounts for this region's greater reduction in cereal area per head (see Table 3.2).

Latin America, North America/Oceania and Europe/FSU

Jointly these three regions produce about half the world's cereals. And it is developments in these regions which really explain the decline in global per capita cereal output since 1984. In each region the 1980s saw a reversal in cereal production per head, after three decades of steady rise. In North America and Latin America the downturn occurred in the early 1980s (see Figure 3.7). In Europe/FSU it was a little later, around 1988.

In all three regions the reversal was the outcome of *both* sharp reductions in total area cultivated with cereals, and a coterminous slackening in the percentage yield growth rate. These changes were most dramatic in traditional exporting countries like the United States and Canada. Table 3.2 shows that comparing averages for 1980–86 and 1987–92, the North America/Oceania region reduced its total area harvested of cereals by over 11 per cent. It also experienced by far the greatest deceleration in yield growth rate, declining from a long-run average of 2.41 per cent per year between 1951–57 and 1980–86, to only 1.13 per cent per year since 1980–86 (an absolute change of −1.28: see Table 3.2). The time series for Latin America perhaps provide the most eloquent example of how simultaneous reductions in per capita area har-

[25] This surge in Chinese cereal yields was also partly due to the development and introduction of hybrid rice. See also the considerations of footnote 2.

vested and yield growth rates have combined to reduce cereal production per head (see Figure 3.8). But, to reiterate, essentially the same processes have been at work in all three of these world regions (see Table 3.2).

RÉSUMÉ AND COMMENTARY

It is obvious from this regional review that the decline in world per capita cereal production since 1984 has little to do with neo-Malthusian processes. The decline has been almost entirely due to changes in North America/Oceania[26] and Europe/FSU – the two world regions with the slowest rates of population growth – plus Latin America. In the Middle East, South Asia and the Far East, recent trends in per capita cereal production have generally been quite good. Only in sub-Saharan Africa can we perhaps detect any Malthusian echoes, and even there it is clear that many other factors are also implicated.

Table 3.2 shows that, in general, those world regions where total area harvested of cereals has expanded since the early 1980s have also tended to experience increases in growth rates of cereal yields.[27] More importantly, the three regions which significantly reduced their total cereal area also experienced the greatest reductions in cereal yield percentage growth rates. This constitutes a powerful suggestion that the explanation for the recent decline in world per capita cereal production lies in reduced incentives for farmers in the world's more developed regions to both plant and grow cereal crops.

Again, the long-run inverse relationship between movements in world cereal stocks and international cereal prices provides an explanatory key (see Figure 3.6). By the late 1970s and early 1980s world cereal stocks had recovered from the crisis of 1972–74 and were already equivalent to over 20 per cent of annual global consumption.[28] At the same time – and because of heavily

[26] Comparing 1980–86 and 1987–92 the rate of change of world per capita cereal production averaged −0.52 per cent per year (see Table 3.2). However, removing only North America/Oceania from the comparison virtually eliminates any decline in per capita output in the rest of the world.

[27] However, in the case of sub-Saharan Africa the increase in cereal yield growth rate was due to recovery from the crisis of the early 1980s.

[28] And, of course, population growth means that this roughly constant proportion represented a rising absolute volume of stocks. It is worth noting too that FAO's recommendation is that the proportion should be equivalent to 17 per cent of annual world consumption.

subsidized prices paid to farmers under its Common Agricultural Policy (CAP) – the European Community emerged on the world stage as a major cereal exporting bloc. By the mid-1980s competition between the major exporters – notably the United States, Canada and the EC (all three of which subsidize their farmers to varying degrees) – was intense. World cereal stocks rose to over 27 per cent of consumption in both 1986 and 1987. Donations of cereal aid rose sharply too – partly to mitigate storage problems. And between 1981 and 1987 the international price of cereals fell by over 40 per cent (see Figure 3.6).

The reactions of the major cereal exporting blocs to these developments have been broadly the same. They have reduced the subsidies paid to farmers to grow cereals, and they have instigated policies to reduce the amount of land sown with cereals. The reactions of the United States and Canada have generally been swifter, and greater, than those of the European Community. But by the late 1980s neither bloc could possibly avoid the fundamental logic of the situation.[29]

The tumble in international cereal prices naturally benefited the many countries which purchase cereal imports. Indeed, because of hidden export subsidies it may be that the true extent of the recent price-fall is actually understated in Figure 3.6. But traditional cereal exporting countries like Australia and Argentina,[30] which cannot afford to subsidize their farmers – and, more importantly, poor cereal farmers in the many countries which have preferred to take advantage of cheap cereal imports, rather than invest in their own production potential – have faced mounting difficulties since the early 1980s.

For the 67 developing countries in our sample Figure 3.9 plots recent growth rates of per capita cereal output against corresponding rates of population growth. Not surprisingly, it provides no support for the proposition that faster rates of population growth are associated with slower rates of per capita cereal production growth. This failure to show a negative relationship is because

[29] A notable step in the United States was the 1985 Food Security Act. These issues are discussed further in chapter 6.

[30] The decline in Latin America's per capita cereal output since about 1983 (see Figure 3.7) is almost entirely due to production cut-backs in Argentina. The country accounted for over a quarter of Latin America's cereal production at the start of the 1980s. But it simply could not afford to compete with the producers of North America and the EC on the prevailing terms.

many different factors – some global, some regional and some country-specific – are combining to influence national trends in cereal output. At one extreme, the fastest growing population, Saudi Arabia (number 33), explicitly decided to invest heavily in wheat production and raised its cereal output massively in the 1980s from a very low base (see also Figure 3.4).[31] At the other extreme, Afghanistan (number 37), the slowest growing population, had both its cereal and demographic growth sharply curtailed by war.

The sample is evenly divided in Figure 3.9 between countries with positive and those with negative per capita cereal production growth rates. And, interestingly, this near-equal division applies within each of the five regions represented. However, we should not become too distracted by 'flags' as opposed to numbers of people.[32] Countries with positive per capita cereal growth (i.e. those above the horizontal line) represent 80 per cent of the total developing country sample population, and include such demographic giants as China, India, Indonesia and Brazil.

Of the 33 developing countries with negative per capita cereal production growth rates in Figure 3.9 one clearly identifiable category are those affected by war. As well as Afghanistan (see also Figure 3.4), this group includes Burundi, Ethiopia, Mozambique, Rwanda, Angola, Somalia, Sudan and Yemen. A second fairly distinct category consists of the small number of cereal exporting developing countries – all of which have been badly squeezed by the plummeting international cereal prices of recent years. As well as Argentina (again, see Figure 3.4), this group includes Myanmar, Turkey and Thailand. A third category comprises countries with relatively high incomes which, especially given the low prevailing

[31] Saudi Arabia did this by drawing heavily on underground aquifers which are ultimately supplied by rainfall from its western mountains. These rates of depletion will not be sustainable in the long run. See Postel (1992, pp. 31–2).

[32] Many statements stress the large number of countries which have experienced declines in per capita food production. For example: 'In 75 countries, less food was produced per person at the end of the 1980s than at the beginning.' See Pinstrup-Andersen (1994a, p. 3); 'Between 1978 and 1989 food production lagged behind population growth in no less than sixty-nine out of 102 developing countries for which data are available.' See Harrison (1992, p. 43). Such statements should never be read without keeping in mind the important proviso that many of these countries, especially in sub-Saharan Africa, have comparatively small populations. Thus in 1990 sub-Saharan Africa contained less than 10 per cent of the world's population but 27 per cent of our sample countries.

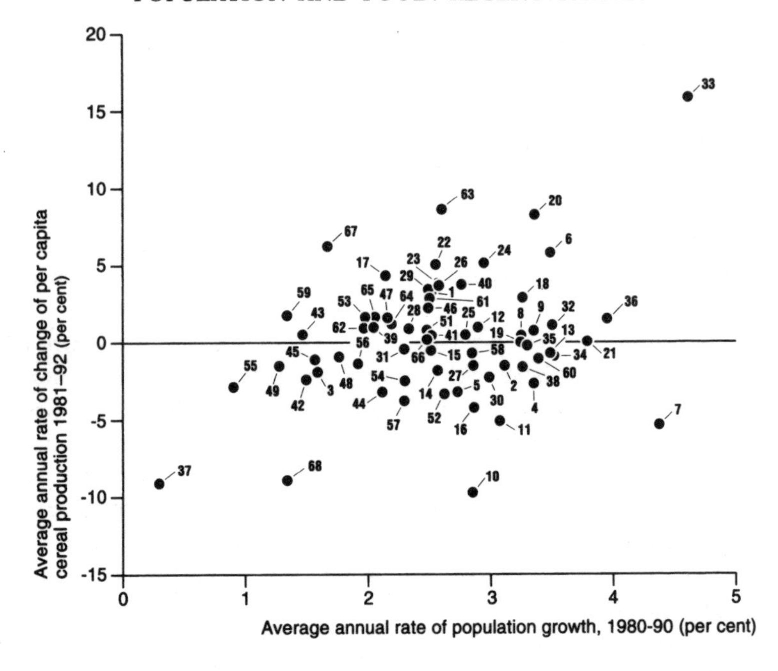

Figure 3.9 Growth rates of per capita cereal production and population, 67 sample developing countries

Note: Country numbers are given in Figure 2.1 and in the Appendix

prices, have increased their cereal imports rather than try to expand their domestic production per head; this group includes Algeria, South Korea, Malaysia and Mexico. Together the countries listed above in these three categories account for half of all the cases below the horizontal line in Figure 3.9. But the influence of cheap supplies of cereals has probably had an adverse effect on per capita cereal production in many countries which lie both below and above the horizontal line.

This interpretation of Figure 3.9 does not mean that we can entirely discount rapid population growth as one limiting factor upon growth rates of per capita cereal production. Indeed, the preceding discussion of the situation in sub-Saharan Africa has already acknowledged that rapid demographic growth in that region almost certainly *is* operating as one such constraint. Rather, the point being made is that to argue that rapid population growth sometimes restricts increases in per capita production on the basis of international cross-sectional data like that in Figure 3.9, any analysis would have to be contrived.[33]

Certainly, the decline in world per capita cereal production of recent years has resulted mainly from the economic consequences of farm subsidies in rich regions, rather than from environmental or demographic trends in the developing world.

CHANGES IN HARVEST VARIABILITY

Another factor fuelling current neo-Malthusian concerns is the intimation of increased variability in the global harvest. And we have already remarked on the occurrence of several major regional droughts during the 1980s and early 1990s.

One way of considering this issue is to examine changes in the extent to which annual per capita cereal production deviates from the five-year moving average curves (i.e. those in Figures 3.1 and 3.7). For the world and its regions Table 3.3 summarizes mean absolute percentage deviations from the respective curves for different time periods. In most decades harvest variability has been greatest in the Middle East – a region with scarce renewable water supplies, where harvests fluctuate greatly according to the weather; Syria presents a typical case (see Figure 3.4). Variability is generally quite modest in South Asia and, still more, the Far East – both regions where major irrigation systems help to restrict harvest instability. Furthermore, harvest variability is quite limited in Latin America and Europe/FSU. For all these regions there is no real evidence of increasing variation; indeed, if anything the 1980–92 period was one of comparatively low harvest variability (see Table 3.3).

But at the global level there may be some signs of increasing variability. Whereas in the 1960s the mean deviation for the world was ±1.43 per cent, by 1980–92 this had risen to ±2.84 per cent (Table 3.3). Sub-Saharan Africa has experienced a continual

[33] The issue being considered is so complex that the same is probably true of any conceivable multivariate extension. It is also worth remarking that, using countries as units of analysis, the plot of per capita calorie or cereal consumption growth against population growth gives equally feeble support for the existence of a negative relationship. As is often the case, the statistical tests and data at hand are probably not up to the job of disentanglement that is required.

Table 3.3 Measures of cereal harvest variability, by world region and period

Region	*1952–59*	*1960–69*	*1970–79*	*1980–92*	*1952–92*
Sub-Saharan Africa	1.08	4.50	4.86	7.69	4.93
Middle East	4.38	6.78	7.92	6.62	6.54
South Asia	3.11	3.83	4.19	2.76	3.43
Far East	2.13	2.58	2.49	1.95	2.27
Latin America	4.10	3.20	3.02	3.81	3.53
Europe/FSU	2.93	3.11	5.23	2.37	3.39
North America/ Oceania	3.84	3.96	5.34	11.15	6.55
World	2.24	1.43	2.42	2.84	2.27

Note: The measures of variation given above are the average absolute percentage deviations from the moving average curves, within each period. Since moving averages are used, the periods for which the measures of deviation can be calculated are slightly truncated. Not surprisingly, the regional measures of variation tend to be greater than the global measures

increase in harvest variability. And, especially in countries located in eastern and southern parts of the region, there are disturbing intimations of greater frequency of drought: witness 1983, 1984 and 1992 (see Figure 3.7). However, this has a negligible impact at the global level because sub-Saharan Africa produces under 5 per cent of world cereals. Instead, the main explanation for the rise in world harvest variability clearly lies in North America/Oceania. This region's measure of variability has also been on a rising trend, and it doubled between 1970–79 and 1980–92 (Table 3.3). Again there have been several recent major droughts – notably in North America in 1980, 1983 and especially 1988. No one can be sure whether these droughts fall within the range which might be expected from previous experience (e.g. the 'dust bowl' years of the 1930s) or whether they are portents of global warming.[34] Certainly, if the three drought years are eliminated from the calculations then no increase in harvest variability is indicated for North America/Oceania between 1970–79 and 1980–92.[35] In addition, the period 1980–92 also saw increased annual variability

[34] See Brown (1988, pp. 5–6).

[35] The years of drought can be eliminated by assuming that the cereal harvests in 1980, 1983 and 1988 were equivalent to the average of the harvests for the two adjacent years; the resulting measures of variability are then changed to 3.96 for 1960–69 (see Table 3.3), 4.91 for 1970–79 and 4.69 for 1980–92.

eal cropland set aside by the United States. And by
cidence the reductions in cereal production due to
1983 and 1988 were compounded by the idling of
ge areas of US cereal land.[36]

...he recent modest increase in global harvest variability is mostly explained by the occurrence of several severe droughts in North America, compounded slightly by increased variation in the areas of cereal cropland set aside there. Variability increased too in sub-Saharan Africa – entirely because of droughts. But in most other world regions the 1980s and early 1990s saw relatively low levels of harvest variability.

TRENDS IN FOOD TRADE AND AID

Recent decades have seen a massive expansion of the world trade in food.[37] Comparing 1950 and 1990 the volume of cereals traded internationally has risen from under 40 million metric tons (i.e. around 6 per cent of global cereal production) to about 225 million tons (i.e. about 12 per cent of production). Many factors have influenced this expansion – including better communications and transport, national and international policies, and changing patterns of food demand.

However, probably the most important reason for the expansion in trade hitherto has been economic growth. Rising incomes – e.g. in parts of the Far East and Middle East – have raised demand for food, including cereals for the production of meat. Such increases in demand have sometimes been so rapid that they could not possibly have been met from increased domestic production. As we saw in chapter 2, around 1990 the former Soviet Union and Japan were the world's largest cereal importers, with China fast establishing itself in third place. And individual countries like Algeria, South Korea and Mexico each imported more cereals than all the countries of sub-Saharan Africa combined.

Measured as a proportion of the total cereal imports of the

[36] For details of annual areas of cropland idled in the United States see Brown (1988, p. 46) and USDA (1992).

[37] For a general review see Grigg (1993, pp. 236–55). He notes that in monetary terms cereals are second only to petroleum in international trade, and that the agricultural exports of the developing world have steadily declined as a proportion of world trade. Clearly too, the long-run decline in agricultural commodity prices (e.g. coffee, tea, sugar, jute) has adversely affected many poor countries and, other things being equal, reduced their capacity to finance food imports.

developing world, food aid has declined sharply since the 1
This decline largely reflects the policy of the United States
increase its sales and reduce its donations – partly in order to finance its oil imports following the events of 1972–75.[38] It is obvious from Figure 3.6 that world food aid volumes vary directly with the size of stocks and inversely with the price level. As well as clouding international cereal prices, the 1980s export clash between the North American producers and the European Community also further confused the meaning of 'food aid', since many 'sales' actually involved hefty subsidies.

Table 3.4 summarizes regional trends in cereal trade and aid as a proportion of regional consumption. Perhaps the most striking feature is that – probably partly because of poor production performance and partly because of the availability of low-priced cereal supplies from overseas – sub-Saharan Africa has become increasingly dependent upon both traded cereal imports and cereal aid.[39] The statistics for the Middle East underscore the huge jump in traded imports after the oil price rises of the 1970s, since when the percentage contribution of imports to consumption has changed little. For South Asia probably the most interesting feature is the declining contribution of cereal aid to total consumption – this is chiefly because India is no longer a major recipient of food aid. From being a major exporter, Latin America has become a significant cereal importing region. In part this reflects the lower volume of cereal exports from Argentina; but most countries in Latin America, including oil producers like Mexico and Venezuela, have increased their net cereal imports during recent decades as prices have fallen.[40] Increased exports from the EC largely explain why the statistics for Europe/FSU show a sharp recent decline in the contribution of traded imports to total consumption.[41] Finally, the dramatic rise in exports shown for North America/Oceania between 1969–71 and 1979–81 largely reflects the adoption of a

[38] In the 1960s 'food aid' comprised 40 to 60 per cent of total US grain exports. By 1990 this had fallen to about 8 per cent. See Fletcher (1992, p. 16).
[39] For an analysis supporting this dual explanation see Jaeger (1992).
[40] However, as chapter 6 elaborates, in the case of Latin America concentration upon cereal transfers gives a somewhat unrepresentative impression of food flows in general.
[41] For example, between 1979–81 and 1989–91 France alone increased its net cereal exports from 17.9 million to 28.8 million tons. Some non-EC European countries (e.g. Austria, Hungary) also sharply improved their net cereal trade balance during the 1980s. See World Resources Institute (1994, p. 299).

Table 3.4 Regional cereal trade and aid expressed as percentage of total regional cereal consumption, 1969–71 to 1990–92

Region	*1969–71*		*1979–81*		*1990–92*	
	Trade	*Aid*	*Trade*	*Aid*	*Trade*	*Aid*
Sub-Saharan Africa	1.5	0.9	4.8	1.8	5.6	3.6
Middle East	10.8	3.4	26.5	3.4	28.0	2.6
South Asia	3.7	2.5	2.2	1.1	3.4	0.9
Far East	6.7	1.0	10.1	0.3	8.4	0.0
Latin America	−6.7	1.1	7.0	0.4	8.6	1.4
Europe/FSU	5.5	−0.2	11.8	−0.1	2.4	−0.1
North America/ Oceania	−30.6	−5.6	−66.4	−3.2	−51.4	−3.1

Notes: (a) Negative numbers denote net exports or donations. Cereal consumption is defined as the volume of regional production, plus net trade and aid flows. Note that the trade and aid percentages shown for North America/Oceania would be markedly smaller if they were expressed in relation to levels of regional production

(b) The figures have been computed on the basis of the sample countries. Due to unavailability, in some cases data for slightly different years to the specified periods were used. Also, totals in the basic data did not always add due to divergences in the statistics between different sources

Principal sources: FAO (various years) *Production Yearbook*, Rome; FAO (various years) *Food Aid in Figures*, Rome; World Resources Institute (1994)

more aggressive food sales policy by the United States. And the reduction in net exports as a proportion of consumption in North America/Oceania between 1979–81 and 1990–92 in part mirrors the increased export competition from the EC.

In conclusion, while the volumes of cereals involved in trade and aid certainly increased during the 1980s, and the role of North America/Oceania as the single major external source of food supplies has been challenged, in most developing regions the decade saw only modest rises in the proportional contribution of imports and aid to total cereal consumption.

TRENDS IN FOOD PRODUCTION, AVAILABILITY AND VARIETY

So far our discussion might be criticized as demonstrating something of a 'cereal mentality'.[42] Cereals are extremely important. But to use a common quip, people do not live 'by bread alone'!

[42] For this expression see Crosson and Anderson (1992, p. 5).

In fact, the long-term rise in world per capita cereal production of recent decades has been mirrored by increases for most other major foods. For example, comparing the early 1950s and the early 1990s global per capita soybean output rose from about 7 to 20 kilograms; and meat production rose from around 18 to 32 kilograms per head.[43] Comparing averages for 1979–81 and 1990–92 world per capita production of root and tuber crops fell by some 13 per cent; but global per capita production of vegetables (inclusive of fruit and nuts), meat and pulses is estimated to have increased by 6, 9 and 13 per cent respectively.[44]

Because 'food' is such a composite category, FAO has developed an index of per capita food production which uses price data to weight production estimates of all the main foodstuffs (i.e. nuts, pulses, fruit, cereals, vegetables, sugars, roots, edible oils, and livestock and livestock products) excluding quantities used for seed or animal feed.[45] Clearly, such a complex index, based on diverse data and sources, should be interpreted with care. But at least the index attempts to capture an increasingly important feature of global patterns of food production and consumption – namely their increasing diversity.

Whereas between 1979–81 and 1990–92 world per capita production of cereals barely increased at all, there was a 5 per cent rise in the FAO index of food production per head (see Table 3.5). Overall, during the 1980s and early 1990s global per capita food production seems to have increased at a similar pace as in the 1970s. Only North America/Oceania and sub-Saharan Africa actually experienced declines in per capita food production between 1979–81 and 1990–92. Latin America and Europe/FSU experienced decelerations in growth compared to the previous period, but in both these regions estimated food output per head still rose slightly. Very significantly, South Asia and the Far East have experienced recent sharp accelerations in this measure of food production. Indeed, the progress achieved by these two Asian regions means that for the developing world as a whole, the index of food production per head rose by 14 per cent between 1979–81 and 1990–92, compared to only a 10 per cent rise between 1969–71 and 1980–82. To the extent that there has

[43] See Brown *et al.* (1993, pp. 28–31).
[44] Based on output statistics from FAO (1993).
[45] For more details see, for example, FAO (1994, p. ix).

Table 3.5 Indices of per capita cereal production, food production and calorie availability, world regions, selected periods

	Cereal production		*Food production*		*Calorie availability*	
Region	*1980–82 (1969–71=100)*	*1990–92 (1979–81=100)*	*1980–82 (1969–71=100)*	*1990–92 (1979–81=100)*	*1979–81 (1969–71=100)*	*1988–90 (1979–81=100)*
Sub-Saharan Africa	93	79	89	96	98	99
Middle East	96	105	101	103	117	106
South Asia	101	108	104	117	103	107
Far East	117	115	118	128	115	110
Latin America	110	91	111	103	108	100
Europe/FSU	109	105	112	103	103	101
North America/Oceania	130	96	116	99	102	108
World	109	101	106	105	106	105

Notes: (a) The regional figures are based on the sample countries. As elsewhere, due to unavailability, in the case of a few countries data for slightly different (usually adjacent) years were used to the specified periods in calculating the above indices. However, the effect on the regional measures is minimal

(b) The FAO indices of per capita food production are calculated by the Laspeyres method, weighting production quantities by prices during the base periods shown. The regional measures have been variously computed using both international commodity prices and populations as weights

(c) As elsewhere, due to changes in data sources, concepts, coverage, weights, etc. over time, the food production and calorie availability measures shown above should be regarded as broadly indicative rather than exact. Also, indexed values are naturally partly dependent upon the particular periods used, a consideration which limits very detailed comparison of the figures

Principal source: FAO (various years) *Production Yearbook*, Rome

been any recent deceleration in the growth of global per capita food production it has clearly been located chiefly in the more developed regions of the world (see Table 3.5) and has essentially stemmed from precisely the same logic as for cereals: namely production cut-backs made necessary by subsidized overproduction in relation to effective demand.

Furthermore, the recent reductions in areas of cropland sown with cereals have sometimes involved transferences of land to other – often higher value – crops. To give a few examples: in parts of Latin America agricultural land has been switched from cereals to soybean cultivation; in Europe the 1980s saw sharp rises in rapeseed, linseed and sunflower production – reflecting increased incentives to plant non-cereal crops; in coastal areas of Bangladesh and India, land once sown with rice has been flooded with sea water in order to cultivate prawns for export; and in parts of northern China cropland has been transferred from cereals to potatoes. Comparing 1980 and 1993 China slightly reduced its total area harvested of cereals. Nevertheless, it managed to raise per capita cereal production and achieved sharp increases in the output of most other foods – like sugar, edible oils, fruit, meat and aquatic products.[46] China's gains largely account for the sizeable 28 per cent rise in food output per head in the Far East region between 1979–81 and 1990–92 (see Table 3.5).

It has long been known that people's food preferences change as their incomes rise.[47] And, as average incomes have risen in many countries since the 1950s, so the structure of demand for food has become much more diverse. With the exception of sub-Saharan Africa, such income-induced changes in demand have been fairly widespread. They greatly inform the recent food output rises in China. And analogous changes are detectable for many other developing countries. In general, these changes in demand stimulate and correspond to the production changes which we have noted above. As incomes rise there tends to be lower relative consumption of rice, increased preference for wheat and its products like noodles and bread, and increased demand for vegetables,

[46] See, for example, State Statistical Bureau of the People's Republic of China (1994).

[47] In 1767 James Steuart wrote: 'The more rich and luxurious a people are the more delicate they become in the manner of living; if they fed on bread formerly they will now feed on meat; if they fed on meat they will now feed on fowl.' Quoted in Blaxter (1986, p. 31).

fruit and meat.[48] Such developments have been influenced too by increased urbanization, communications and changing employment patterns. Thus, to feed growing urban populations, cereals are probably easier to transport and store than are bulky root crops; and the growth of mass communications has probably promoted tastes for western wheat-based products – which may also be easier and quicker to prepare. Even in sub-Saharan Africa there are strong indications that fast-growing urban populations – many of which are also coastal ports – now often prefer imported wheat from overseas rather than indigenous staple foods.[49]

The indices of per capita calorie availability given in Table 3.5 provide a partial summary of the outcomes of these various developments – in food production, diversity, and flows of trade and aid. They too indicate little difference in the overall rate of improvement between the 1970s and 1980s. The food production decline in sub-Saharan Africa has evidently been moderated by increased imports and aid, and estimated calorie availability in the region has fallen only slightly. Latin America's recent performance has broadly been one of stagnation – largely due to a poor production record, compounded by the region's debt crisis and restricted consumer demand. But all other world regions are thought to have experienced increases in average per capita calorie supply during the 1980s. The most important gains were in the Far East and South Asia, and were clearly underpinned by significant advances in food production.

CONCLUDING REMARKS

The most important general conclusion of this chapter is that the neo-Malthusian representation of recent world food production trends is largely mistaken. It is true that since about 1984 world population growth has outpaced cereal production. It is true that

[48] For a fine review of these changes see Mitchell and Ingco (1993, ch. V). It is worth remarking that increasing diversity of human food consumption has probably been mirrored in patterns of animal feed. For example, the recent increase in feeding manioc to livestock in South-east Asia provides an example of a relative shift away from cereals for feed.

[49] Of course, some of these changes in food consumption may be undesirable. For example, relatively healthy traditional diets involving a good mix of pulses, vegetables and cereals can be swiftly disrupted by urban life. Also, there is abundant evidence that customary and sound breastfeeding patterns are being disrupted faster in urban than in rural areas of many poor countries.

the global harvest may recently have become slightly more variable. And sub-Saharan Africa is manifestly having difficulty in expanding its food production to match its demographic growth. But these three are really the only points which can be rescued from the neo-Malthusian portrayal of recent trends; and the import of the first point is debatable.

Conversely, it is *not* true that population growth has been out-pacing cereal production in all main world regions.[50] On the contrary, recent trends in per capita cereal production in the two largest developing regions have been positive; the decline in world per capita cereal output since 1984 is explained by policy changes in the most developed regions – which also have the slowest growing populations.

It is *not* true that world cereal yields are 'plateauing' in the sense that they are levelling-off. Since the early 1980s cereal yields have continued to rise in all world regions. Measured – as it probably should be – on a linear scale, the average global cereal yield is continuing to rise with reasonable momentum. Because of the linear nature of this rise it is to be expected that the annual percentage increment will tend to decline. However, even if we examine yields in terms of average annual percentage growth rates it is clear that any slowdown in growth has mainly been confined to the more developed world regions.

In a majority of the seven regions used in this study cereal harvest variability actually declined between 1970–79 and 1980–91. The risk and severity of famines is *not* increasing, and their numerical demographic impact has lessened in recent decades.

Finally, the 1980s and early 1990s have *not* been a particularly dismal period for global per capita food production. Given factors like rising incomes, urbanization and expanded trade, patterns of food production and consumption are becoming increasingly diverse; and though still very important, cereals may be a less adequate proxy for general food trends. Indeed, unlike per capita cereal production, there has been no recent deceleration in the rate

[50] See chapter 1 (p. 17) for a statement by Lester Brown which implies that per capita cereal production downturns have occurred in all main world regions. Brown's substantiation of this position is based on showing that per capita output in a single year (1990) was lower than in some previous year (e.g. 1984 in Asia). But several of the resulting 'downturns' therefore reflect annual harvest variability, rather than a reversal in the direction of the basic trend. See Dyson (1994a, pp. 399–400).

of food production per head – whether measured in monetary or calorific terms. Even the FAO considers that despite an addition to the world's population of about 841 million between 1980 and 1990, the number of undernourished people probably fell – albeit modestly – from 844 million to 786 million during the decade.[51]

This chapter's analysis supports several other key conclusions. First, it is clear that yield growth – rather than area expansion – has been the dominant proximate cause of increased food output since the mid-1950s. All the signs are that this will be even more true in the years ahead.

Second, however tempting, it is probably unhelpful to conceptualize modern agricultural progress in terms of a 'green revolution' and similar related expressions which are commonly used – like 'miracle varieties' of plants. For sure, the development of HYVs has been extremely important. And it is sometimes possible to detect periods of faster yield growth if data are disaggregated by crop or country. But, in general, food production trends are better conceptualized in terms of complex processes. One implication of this is that in the future we should be wary of expecting too much from any single source, such as the seeming wonders promised by genetic engineering and biotechnology.

Third, the basic relationships linking levels of global food stocks, international food prices and food aid donations during recent decades are extremely clear. Volumes of world cereal stocks and cereal aid donations have varied directly with each other, and inversely with prices. The logic of these relationships is easy to understand. We can be fairly sure that, to some degree, these relationships will hold in the future – although the context in which they operate may be modified by the consequences of the recent subsidized competition between the traditional food exporters of North America/Oceania and the European Union.

Fourth, the risk of a future 'world food crisis' – in the sense of a repeat of 1972–74 – lies chiefly (although not exclusively) in the coincidence of deliberate policy decisions to cut-back food production and run-down stocks, plus the simultaneous occurrence of widespread droughts around the world. A repeat episode of droughts is a virtual certainty at some time. And policy mistakes are inevitable. In recent decades reserve cereal stocks held particularly in the United States (which, importantly, also has the capacity to rapidly expand its planted area) have provided a measure of

[51] See FAO (1992, p. 9).

global food security, mainly as a happy by-product of US domestic farm policy. Experience since the early 1980s suggests that future responsibility for maintaining reserve stocks and disbursing food aid may be more widely spread between nations – a situation with both potential advantages and disadvantages. However, provided stocks are kept at broadly previous levels, the key element in any future crisis will be a sudden rise in prices rather than any real threat of a complete depletion of global food stocks.[52]

Fifth, the recent general decline in international grain prices – which is part of a much longer term declining trend – has generally been to the advantage of net consumers and the disadvantage of net producers. Moreover, almost all developing countries, and many individuals (e.g. people living in urban areas), are in the former category rather than the latter.[53] It would be folly not to acknowledge that one recent indirect outcome of agricultural subsidies in the rich nations has been to benefit many poor consumers. And conversely, the future reduction of these subsidies will hurt many poor consumers. This said, it is also clear that a sizeable fraction of humanity has benefited very little from the recent fall in international prices, either because they are cereal producers (a category which includes many poor farmers) or because they do not interact much with markets.

Finally, we return to yields, because of their critical importance for the future. Like most time series which fluctuate considerably from year to year, yield trends can be represented in very different ways – for example, depending upon which specific years are chosen for comparison.[54] It is clear that yields in

[52] In Figure 3.6 the lowest stock level as a proportion of world consumption occurs in 1973 at 15.1 per cent. The greatest single annual reduction in stocks (−5.5 per cent) occurs between 1988 and 1989, and was deliberately brought about to reduce the level of stocks. During the crisis of 1972–74 the greatest reduction (−3.6 per cent) occurred between 1972 and 1973. These figures suggest that global stocks have been sufficient to cope with at least two consecutive very bad years.

[53] Of the 67 sample developing countries only a few are net cereal exporters (see the Appendix). However, in many of these countries a high fraction of the labour force is employed in agriculture.

[54] For example, Lester Brown uses the year 1984 to illustrate a 'dramatic slowdown' in recent world cereal yield growth. See Brown (1994a, p. 185) and Brown and Kane (1995, pp. 136–42). Indeed, using the FAO cereal data it is true that there was a 2.3 per cent annual increase in global cereal yields between the specific years 1951 and 1984, and only a 1.4 per cent annual increase between 1984 and 1992. However, if the year of comparison is changed to 1983 then cereal yields actually grew at an almost identical average annual percentage rate between 1983 and 1992 (2.16 per cent) compared to the period 1951–83 (2.14 per cent).

sub-Saharan Africa are both inadequate and in trouble – with generally low and even negative percentage growth rates comparing averages over the short to medium run. But elsewhere, by far the most important recent slowdowns in percentage yield growth have occurred in North America/Oceania and Europe/FSU (plus Argentina in Latin America). And these are precisely the same regions which have also had to substantially reduce their planted area because of the prevailing international price conditions. Throughout the rest of the developing world recent average annual yield growth rates have more than offset declines in per capita harvested area. And in most regions there are no obvious signs for alarm regarding recent yield trends. We will return to this issue in chapter 4.

4

EXPLORING THE FUTURE: DEMAND AND SUPPLY

INTRODUCTION

This chapter begins our consideration of future prospects. Building upon the analysis of past trends in chapter 3, the principal objective here is to use some simple assumptions and calculations to explore the possible evolution of cereal demand and supply to the year 2020. The chapter deals with some of the basic 'numbers' – for example, of people, cropland and yields – which are likely to be involved, and tries to assess their general implications for future global and regional patterns of food production and inter-regional transfers of food.

To chart a plausible future scenario the exercise combines elements of projection, extrapolation and judgement. But it is certainly not a forecast, nor, still less, is it a prediction. Rather, we try to sketch the bare bones of one broadly credible future – from which reality will certainly diverge.

The simple and speculative nature of the exercise is reflected in several ways. Inevitably, we use cereals as a crude proxy for food in general. The analysis is conducted mostly at the regional level, although it is important to appreciate that the circumstances of particular countries will sometimes significantly influence the emerging regional picture. We use a plain 'demand and supply' framework, but in reality demand and supply will inevitably interact. In projecting 'demand' it will become apparent that here we are roughly projecting future levels of consumption. Lastly, to emphasize that we are concerned with broad orders of magnitude, many of the numbers referring to the future have been heavily rounded.

Chapters 5 and 6 will provide a chance for some qualification and embellishment! But here we paint with a very broad brush. We address the future scenario first through a consideration of 'demand', second through a consideration of 'supply' and finally through a reconciliation of them both.

FUTURE DEMAND FOR CEREALS

It is important to stress at the outset that in projecting future levels of cereal 'demand' we are not concerned with the difficult and contentious issue of projecting nutritional or food requirements. Rather, future demand will be defined with reference to current levels of (cereal) consumption – irrespective of how inadequate, or excessive, these levels may be.

Table 4.1 provides estimates of future population growth, income growth and urbanization – all factors which are likely to have an influence on the evolution of cereal demand in the next few decades. However, of these three factors demographic increase is likely to be by far the most important. Between 1950 and 1990 roughly two-thirds of the total increase in world cereal production (and hence consumption) can be attributed to population growth.[1] And all the indications are that population growth will be an even more predominant determinant in the future. Therefore we will consider the demographic basis of future cereal demand growth first.

The demographic basis of growth in demand

The 'medium variant' population projections of the United Nations suggest that between 1990 and 2020 the human population will rise by 49 per cent, from around 5285 million to 7888 million (Table 4.1). Probably the most stunning regional projection is that for sub-Saharan Africa, where an increment of 124 per cent is expected, i.e. an addition of 607 million people – the largest proportional and second largest absolute increase of any region. It is worth remarking that this projection includes some – albeit tentative and perhaps too small – allowance for slower population

[1] Between 1950 and 1990 world population grew by 110 per cent; and comparing averages for 1948–52 and 1988–92, world cereal production rose by 174 per cent. Therefore, roughly 63 per cent (i.e. (110/174) × 100) of the increase in cereal production can be attributed to population growth. This procedure has been used to generate the decadal estimates shown in Figure 4.1 and it forms the basis of the ratios given in Table 4.2.

Table 4.1 Factors affecting future demand for cereals, world regions, 1990–2020

Region	*Population (million)* 'Medium' projection		*% change*	*Per capita income (1990, $US)*	*Forecast GDP per capita growth rate (% per year)*	*Urbanization (% urban)*	
	1990	*2020*		*1990*	*1992–2002*	*1990*	*2020*
Sub-Saharan Africa	490	1097	124	480	0.6	31.0	52.1
Middle East	276	511	85	1536	1.6	53.4	70.9
South Asia	1193	1996	67	444	3.4	27.3	47.7
Far East	1794	2387	33	2380	5.9	37.0	61.4
Latin America	440	676	54	2148	2.1	71.5	82.9
Europe/FSU	788	823	4	11,719	—	70.7	80.3
North America/Oceania	304	397	31	21,391	—	74.8	82.6
World	5285	7888	49	4180	—	45.2	62.0

Note: The World Bank's 'baseline' GDP growth rates which are given above relate to slightly different geographical regions from those used in this study. For example, the growth rate for sub-Saharan Africa excludes South Africa; that for East Asia and the Pacific (here shown for the Far East region) excludes Japan. However, for present purposes such differences are not important

Principal sources: United Nations (1991, 1994); World Bank (1992, 1993a); World Resources Institute (1992)

growth due to the HIV/AIDS epidemic.[2] The population of the Middle East is projected to nearly double by the year 2020, a net addition of about 235 million. The third largest proportional rise (67 per cent) is forecast for South Asia, where an addition of some 803 million people constitutes the largest absolute gain envisaged for any region. The proportional increases projected for the Far East and Latin America (respectively 33 and 54 per cent) are smaller; but the corresponding additions of 593 million and 236 million people are still very considerable. Finally, both Europe/FSU and North America/Oceania are projected to experience comparatively modest demographic increases (see Table 4.1).

At the world level these UN projections will probably turn out to be fairly accurate. This is partly because one very important determining factor of future demographic growth – namely the size and shape of the baseline (i.e. 1990) regional population age pyramids – is already known with reasonable precision. In turn, this means that we have a fairly good idea of the numbers of women who will be of childbearing age in the next few decades. What is more difficult to gauge is just how quickly these women – particularly those living in the developing world – will reduce their average levels of fertility.[3] However, the UN has an impressive record of foresight regarding global population growth. For example, in 1957 it projected that in the year 2000 humanity would number 6280 million; and its forecast for that same year made in 1994 is very similar – at 6158 million.[4]

Of course, at the country level the error-margins of population projections are greater – chiefly because of the difficulties in estimating the future speed of national fertility declines. However, at higher levels of aggregation such errors are frequently offsetting, so the projected regional population totals can also probably be regarded as broadly reliable (though hardly exact).

Nevertheless, partly to emphasize the element of uncertainty which is inherent to any such exercise, the UN also publishes 'high' and 'low' variant projections corresponding to assumptions

[2] The incorporation of assumptions relating to the impact of HIV/AIDS in sub-Saharan Africa dates from the 1992 UN world population revision. See United Nations (1993).

[3] By 'fertility' we mean the average number of live births per woman. See Table 2.2 for regional estimates of total fertility per woman.

[4] For the 1957 projection and an illuminating discussion see Lee (1991). For the 1994 projection see United Nations (1994). See also Demeny (1994, p. 21).

that levels of fertility will decline slower, or faster, than in the 'medium' forecast. For the year 2020 these high and low variants put the global population at 8392 million and 7372 million respectively – giving an average range of variation around the medium of ±6.4 per cent. The corresponding ranges for all world regions are similar to this figure.[5]

While, of course, the United Nations' projections take account of regional differences in levels of mortality, they do not incorporate any explicit neo-Malthusian feedbacks via raised death rates – such as were mooted in chapter 1.[6] We have seen that nowadays mortality trends are determined by many factors in addition to food availability *per se*, and that famines are generally on the decline. Furthermore, the momentum of demographic growth in the contemporary developing world is such that even an unprecedented rise in death rates due to famine would soon be made up in narrow numerical terms.[7] So while it would be foolhardy to entirely exclude the possibility of such feedbacks eventually operating to some extent, in certain specific locations, and perhaps particularly over the longer run, it nevertheless seems highly unlikely that neo-Malthusian mechanisms will have much relevance at the regional and global levels between now and the year 2020.

Future interactions between demographic growth and food demand will be more complex and subtle than just the matter of population increase. For example, especially in the Far East region the coming decades will see sharply ageing populations which, other things being equal, may marginally raise the overall volume of demand. Likewise, demographic growth can indirectly influence

[5] In fact these ranges of variation – which should not be confused with probability confidence intervals – are not exactly symmetrical. But little violence is done by portraying them in this way. The regional ranges vary between a low of ±5.5 per cent for sub-Saharan Africa to a high of ±7.8 per cent for North America/Oceania where the volume of future immigration represents a significant additional unknown quantity for projection.

[6] For recent regional measures of mortality see Table 2.2. Because it can be shown that mortality change has an increasingly small influence on the rate of population growth during the course of the demographic transition, the UN makes only one set of assumptions regarding the future course of mortality. These envisage that by the period 2015–20 average life expectancy will be about 62 years in sub-Saharan Africa, and over 70 years in all other regions.

[7] For example, in the early 1990s sub-Saharan Africa's population was growing at around 16 million people per year. A regional population loss of 8 million – for which even the events in Rwanda in 1994 (which may have caused 1 million excess deaths) provide no real precedent – would thus be made up in six months. On this issue see Menken and Campbell (1992) and Watkins and Menken (1985).

the quantum of food demand through its effects on national income distributions.

But these are comparatively minor considerations. Provided we bear in mind that the UN medium variant projections could be out by about 6 per cent either way, they probably provide a fairly sound basis from which to assess the future influence of demographic factors on the growth of food demand.

The non-demographic basis of growth in demand

Economic and social trends cannot be forecast with a similar degree of confidence. And their likely influence on the total volume of food demand is both weaker and more difficult to specify.

The World Bank's 'baseline' economic projections foresee very fast economic growth in the Far East region by the year 2020 (see Table 4.1). This is largely based upon the expectation that China's recent rapid progress will continue and, indeed, that the country may become the centre of a fourth major world economic 'growth pole'. However, the World Bank considers that future income growth in Latin America, South Asia and the Middle East will be much more modest by comparison. And the economic outlook for sub-Saharan Africa is viewed as especially problematic and dismal – much of any 'progress' there essentially consisting of recovery from the disastrous set-backs of the 1980s and early 1990s (see Table 4.1).[8]

Turning to urbanization – which is probably of only residual significance for present purposes – the United Nations forecasts greater urban living in all world regions. By the year 2000 it is likely that a majority of humanity will live in towns; and by the year 2020 this could well apply in every world region. The greatest increases in urbanization are expected to occur in sub-Saharan Africa, the Middle East, South Asia and the Far East. It is noteworthy that if the projections of urbanization in Table 4.1 prove to be realistic, then the size of the world's rural population will remain roughly constant over the period 1990–2020. Because of rural to urban migration, and the natural increase of urban populations, virtually all of the projected global population increase will occur in urban areas.

[8] See World Bank (1993a, 1993b).

Table 4.2 Measures of cereal consumption growth, world regions, 1970–90

Region	*Total cereal consumption increase (%)*			*Ratio of population increase (%) to total cereal consumption increase (%)*			*Average annual rate of change (%) of per capita cereal consumption*		
	1970–80	*1980–90*	*1970–90*	*1970–80*	*1980–90*	*1970–90*	*1970–80*	*1980–90*	*1970–90*
Sub-Saharan Africa	26.8	27.0	61.0	1.22	1.29	1.30	−0.46	−0.61	−0.53
Middle East	52.9	37.1	109.7	0.59	0.88	0.67	1.55	0.34	0.94
South Asia	22.7	39.4	71.1	1.13	0.65	0.81	−0.24	1.05	0.40
Far East	43.8	30.6	87.9	0.47	0.55	0.47	1.76	1.12	1.44
Latin America	51.0	13.9	72.0	0.52	1.64	0.77	1.75	−0.76	0.50
Europe/FSU	18.0	7.2	26.6	0.37	0.74	0.47	1.01	0.17	0.59
North America/ Oceania	13.5	6.4	20.7	0.87	1.64	1.13	0.15	−0.38	−0.11
World	29.2	19.7	54.7	0.70	0.97	0.79	0.71	0.05	0.38

Notes: (a) Total regional cereal 'consumption' here is the outcome of production and net transfers. As elsewhere, figures should be regarded as broadly indicative of trends, rather than exact. For the world as a whole, consumption equals production. At both the world and regional levels, consumption subsumes changes in stocks. The total cereal consumption increases (per cent) shown above are based on three year averages for sample countries (i.e. 1970 = 1969–71; 1980 = 1979–81; 1990 = 1989–91)

(b) See footnote 1 for an illustration of the procedure used to calculate the ratios of population increase to total cereal consumption increase

Principal sources: FAO (various years) *Production Yearbook*, Rome; FAO (various years) *Food Aid in Figures*, Rome; United Nations (1994); World Resources Institute (1994)

Future demand: a synthesis

Besides suggesting that the Far East region may experience considerable economic progress in the coming years (an assessment which itself is essentially just an extrapolation from the recent past) the estimates of income and urbanization in Table 4.1 constitute a feeble basis from which to gauge the influence of non-demographic factors on the future growth of cereal demand. Accordingly, our assumptions on this issue will be based on a consideration of recent experience.

Figure 4.1 provides a valuable, if rough, indication of the extent to which population growth has been responsible for the growth in world cereal consumption during recent decades. It plots the ratio of the percentage increase in world population to the percentage increase in global cereal consumption, within each decade. It is clear that world population growth is accounting for a rising share of cereal consumption growth. Between 1950 and 1960 about 60 per cent of the increase in global cereal consumption can be attributed to demographic growth. This fraction rose to about

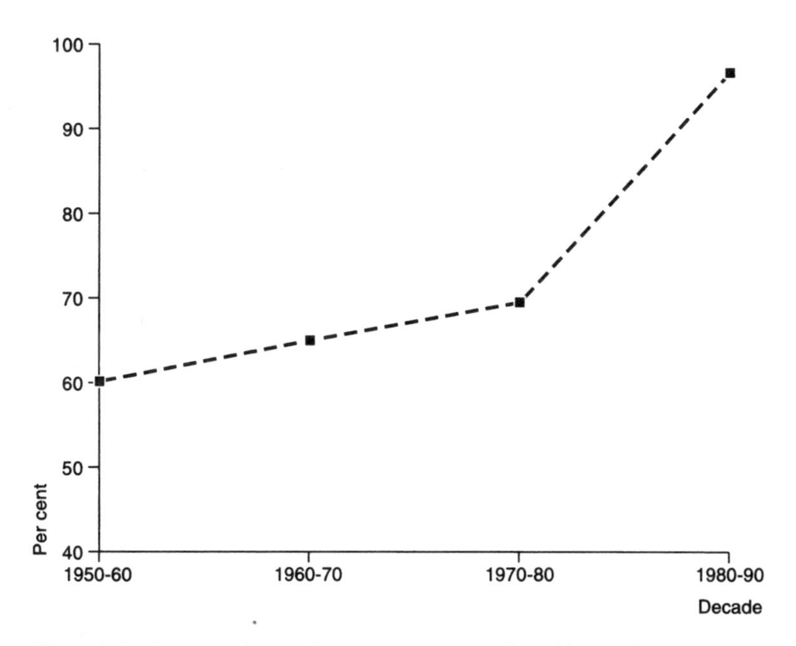

Figure 4.1 Estimated contribution (per cent) of world population growth to world cereal consumption growth, decades 1950–60 to 1980–90

70 per cent for the period 1970–80. For the decade 1980–90 the ratio rises to 97 per cent – implying that virtually all recent growth in cereal consumption can be accounted for by population growth.[9] This last figure is probably something of an exaggeration – being well above the run of the previous trend (see Figure 4.1) – and is doubtless biased upwards by the policy-induced cut-backs in cereal production of recent years. Nevertheless, the central point is very clear: we can expect that population growth will be an even more dominant determinant of the growth in world cereal demand in the period to the year 2020.

Table 4.2 provides similar ratios for each region, together with corresponding average annual rates of change of per capita cereal consumption. The latter, of course, reflect the specific influence of all non-demographic factors on past demand growth. In sub-Saharan Africa population growth outpaced total cereal consumption growth during both the 1970s and 1980s (i.e. the ratios are above unity) and per capita cereal consumption declined. In the Middle East during the 1970s population growth accounted for only around 59 per cent of total consumption growth – per capita cereal consumption rising briskly along with rising oil revenues. However, in the 1980s most (88 per cent) of the Middle East's expanded cereal consumption was due to population increase, and the growth rate of per capita consumption fell sharply to only 0.34 per cent per year. For South Asia, with its somewhat improved economic and cereal production performance during the 1980s, there are indications that non-demographic factors may be accounting for an increasing share of the rise in total regional cereal consumption. Thus during 1980–90 population increase accounted for just 65 per cent of the rise in total consumption. However, we are probably on firmer ground merely to note that over the full period 1970–90 per capita cereal consumption in South Asia grew at only 0.4 per cent per year.[10]

In the Far East during both 1970–80 and 1980–90 the growth in total cereal consumption can be fairly equally attributed to demo-

[9] These percentages were derived by the method illustrated in footnote 1. Recall that 'consumption' here is broadly defined and extends beyond cereals actually eaten by people.

[10] In general the statistics for 1970–90 in Table 4.2 are more robust because they relate to a longer time period. South Asia's faster population than cereal consumption growth during 1970–80 is partly explained by rather low regional cereal production in 1979–81. In turn, this tends to reduce the ratio (0.65) shown for 1980–90.

graphic and non-demographic factors. This is partly evidence of the region's general economic dynamism. However, the Far East's demographic growth rate also slowed considerably over the period, as did the growth rate of per capita cereal consumption (see Table 4.2). The statistics shown for Latin America are more difficult to interpret because of the dramatic slowdown in cereal production – and hence consumption – during 1980–90.[11] The 1980s were certainly particularly difficult years in Latin America, partly because of the region's huge volume of external debt. Again, we are probably on firmer ground in noting that for the period 1970–90 as a whole per capita cereal consumption grew at an average annual rate of 0.5 per cent per year.

Finally Table 4.2 shows that Europe/FSU and North America/ Oceania experienced comparatively small and declining percentage increases in total cereal consumption during 1970–90; and in both regions the contribution of population growth to total cereal consumption growth increased. However, these statistics should be interpreted with particular care because of the strong influence of the recent policy-induced production cut-backs, and because the indicated changes in 'consumption' certainly partly reflect changing levels of cereal stocks. So for these developed regions, where the levels of per capita cereal consumption are already extremely high, perhaps the safest conclusion which can be drawn is that during the 1980s per capita cereal consumption either grew very slowly or actually declined (see Table 4.2).

With this as background, Table 4.3 summarizes projected levels of total cereal consumption/demand in the year 2020 on the basis of several assumptions.

First, for each region, total cereal demand has been projected on the assumption that levels of per capita cereal consumption remain *constant* at the levels already achieved by 1990. In other words, the population growth of the UN medium projections is initially assumed to be the *sole* factor determining the future growth of total regional cereal demand (see the 'population increase only' column in Table 4.3).

For sub-Saharan Africa, Europe/FSU and North America/ Oceania, this assumption alone will probably suffice to express the volume of future cereal demand. In the case of sub-Saharan

[11] Obviously Argentina has been especially significant in shaping this regional picture.

Africa the rationale for this is that, given the generally deteriorating food production situation of 1970–90, the region may do well just to maintain its cereal 'demand' at levels corresponding to 1990 levels of per capita consumption. Moreover, we have seen that the economic conditions of this region are not expected to improve much.

In the cases of Europe/FSU and North America/Oceania the same assumption is justified, but by a very different rationale. In these regions current levels of total (i.e. direct plus indirect) cereal consumption are already so high that it seems reasonable to suppose that demand is effectively saturated. So, for example, any future rises in per capita incomes in these regions are unlikely to create much additional demand for cereals.[12] Of course, it is certainly possible that levels of per capita cereal consumption in Europe/FSU will climb towards the even higher levels prevailing in North America/Oceania; and even a modest move in this direction would exert a sizeable influence upon the region's projected volume of total cereal demand.[13] But it is probably just as likely that in both these world regions the future trend in cereal consumption per head will be *downwards* – as people shift to less meaty diets. This will probably happen more for health than for economic reasons.[14]

However in the remaining world regions it does seem appropriate to make some additional allowance for the impact of non-demographic factors on the growth of cereal demand. This has been done with reference to the past average annual rates of per capita consumption change already reviewed in Table 4.2.

For the Middle East we have simply assumed that the much slower growth rate of per capita cereal consumption experienced during the 1980–90 decade will continue – leading to a level of per capita cereal consumption of around 427 kg by the year 2020 (see Table 4.3). For reasons on which we have already touched, in the

[12] Crosson and Anderson (1992, p. 9) describe current levels of per capita cereal consumption in most developed countries as being 'close to the biologic limit'.
[13] To illustrate the point: if by the year 2020 Europe/FSU were to experience the average levels of per capita cereal consumption 'enjoyed' by the populations of North America/Oceania around 1990, then the region's projected volume of total cereal demand would be 642 million tons, rather than the 521 million shown in Table 4.3. Note that this additional increase in annual demand of 121 million tons would be much larger than the total increase being projected for sub-Saharan Africa in the same table.
[14] On this possibility see Popkin (1993).

Table 4.3 Projected demand/consumption of cereals, world regions, 2020

Region	*Cereal consumption around 1990*		*Projected total cereal demand (million mt) based on:*			*Per capita cereal consumption in 2020 (kg)*	*% change of per capita consumption 1990–2020*	*% of total world demand increase 1990–2020*
	Per capita (kg)	*Volume (million mt)*	*Population increase only*	*Population increase plus non-demographic factors*	*Total*			
Sub-Saharan Africa	150	75.4	164.8	—	164.8	150	0	9
Middle East	386	105.2	197.4	218.4	218.4	427	10.6	11
South Asia	237	281.8	472.2	532.6	532.6	267	12.7	24
Far East	338	606.0	806.1	1004.9 to (1128.8)	1004.9	421 to (473)	24.6 to (39.9)	38
Latin America	265	116.9	179.3	208.1	208.1	308	16.2	9
Europe/FSU	634	500.8	521.6	—	521.6	634	0	2
North America/Oceania	780	236.6	309.3	—	309.3	780	0	7
World	363	1922.9	2650.8	—	2959.7	375	3.3	100

Notes: (a) The per capita cereal consumption figures for 1990 are averages for 1989–91 based on the sample countries within each region. When combined with the corresponding regional population totals for 1990, these figures give the corresponding estimated volumes of total regional cereal consumption

(b) To illustrate the calculation of the lower (preferred) level of total cereal demand in the Far East based on population increase plus non-demographic factors, it was assumed that, within each decade, total demand rose by a factor of 1.82 (i.e. 1.0/0.55) times the demand increase indicated by demographic growth alone. For the factor 0.55 representing the contribution of population increase to total demand increase during 1980–90 see Table 4.2. Compounded across three decades, this procedure actually leads to a slightly lower (51 per cent) ratio of population increase to total cereal demand growth

(c) Note that the lower level of projected demand for the Far East has been assumed in calculating the total and per capita world consumption figures for 2020, and the corresponding percentage changes

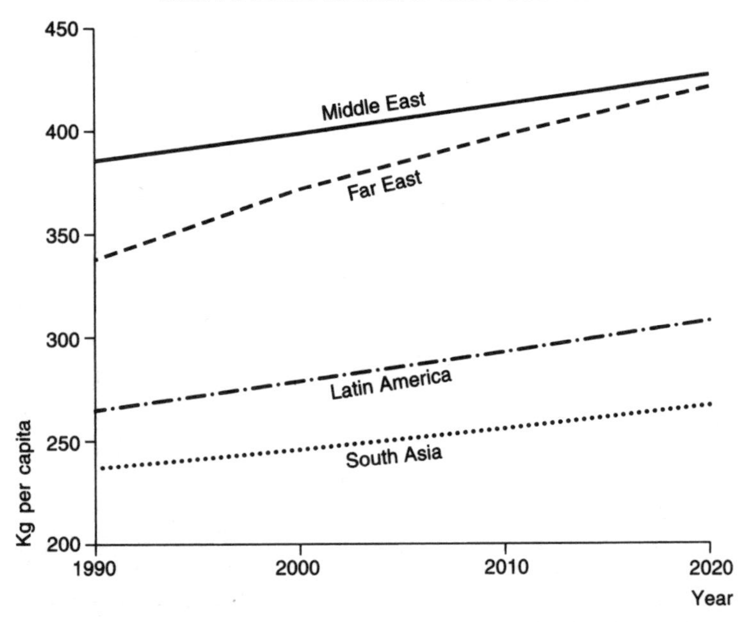

Figure 4.2 Assumed trajectories of per capita cereal consumption, 1990–2020

Note: The preferred lower trajectory is shown for the Far East

cases of both South Asia and Latin America it seems advisable to place most confidence on the average annual growth rates of per capita cereal consumption experienced for the full twenty-year period, 1970–90. The continuation of these growth rates would result in per capita cereal consumption levels in the year 2020 of 267 kg and 308 kg respectively (Table 4.3). Finally, in the case of the Far East the continuation of the slower 1980–90 growth rate would result in per capita consumption of 473 kg in the year 2020. But in view of the indication that this region's growth rate of per capita cereal consumption is already slowing, a second preferred projection has been made on the basis that within each decade of the period 1990–2020 demographic increase will account for 55 per cent of total cereal demand growth.[15] This favoured alternative assumption results in per capita consumption of about 421 kg by the year 2020 (Table 4.3) and reflects quite well the idea that growth in cereal demand may slacken at higher levels of consump-

[15] See the notes to Table 4.3.

tion. Figure 4.2 summarizes the outcome of these assumptions for these four developing regions.

Several interesting indications emerge from the results of this exercise (see Table 4.3). First, if the growth in global cereal demand is projected simply on the basis of population increase (i.e. with regional per capita consumption levels held constant) then the total volume of cereal demand will rise by 38 per cent, from around 1.92 billion to about 2.65 billion tons. This percentage rise is appreciably *less* than the corresponding 49 per cent increase in world population (see Table 4.1) because the regions where most of the projected demographic increase will occur have lower-than-average levels of per capita cereal consumption. Therefore, just because of this 'composition effect' world per capita cereal consumption would *fall* by about 7 per cent.[16]

Second, on the present assumptions, if some allowance is made for the influence of non-demographic factors on future demand growth in the Middle East, South Asia, the Far East and Latin America, then the total volume of world cereal demand in 2020 will rise to about 3.0 billion tons (the precise figure depends upon which estimate of demand is selected for the Far East). In turn, this means that roughly 80 to 90 per cent of the increase in global cereal demand by the year 2020 will be attributable to population growth.[17] Note that such a range is reasonably consistent with the general trends suggested by Figure 4.1.

Third, it is clear that the lion's share of the future growth in global cereal demand will occur in the Far East (especially China). Even on the basis of the lower figure of projected consumption, this region will account for some 38 per cent of the total increase in world cereal demand (see Table 4.3). And both demographic and non-demographic factors are expected to be important determinants of demand growth in this region. Almost certainly South Asia will be the next most important contributor (24 per cent) – although there the main determining factor will be population

[16] If global cereal consumption in the year 2020 were 2651 million tons and the world population were 7888 million, then the average per capita level of cereal consumption would be only 336 kg.

[17] The projected world population increase is 49 per cent (see Table 4.1). The projected rises in total cereal consumption are 54 per cent (i.e. 2959.7/1922.9) and 60 per cent (i.e. 3083.6/1922.9) – depending upon which consumption figure is taken for the Far East (see Table 4.3). Therefore the corresponding range is 91 per cent (i.e. 49/54) to 82 per cent (i.e. 49/60).

growth. In comparative terms, all other world regions will surely play much smaller roles.

Fourth, if these assumptions have broad validity then the greatest increases in per capita consumption will occur in the Far East. By the year 2020 cereal intake per head in this region will either rival, or exceed, the levels projected for the Middle East (see Figure 4.2). This said, if only the lower level of average consumption (i.e. 421 kg) is attained (which we suspect is more plausible) then in absolute terms this will be a smaller rise than was achieved in the Far East during the twenty-year period 1970–90. And even should the higher level (473 kg) be attained then the region's average level of per capita cereal consumption in 2020 will still be lower than was achieved by North Korea around 1990.[18]

Finally, Table 4.3 illustrates very well how significant regional gains in per capita consumption may easily be obscured in the global statistics. After all, appreciable rises in average per capita cereal intake are envisaged for four regions. But the corresponding increase at the world level is a mere 3.3 per cent.[19] This is eloquent testimony to how misleading it can be to fuss overly about the global statistics!

To sum up, the 'broad-brush' nature of these assumptions and illustrative calculations regarding future cereal demand should require no emphasis. Recall that even the demographic projections have a significant margin of uncertainty attached. Nevertheless, Table 4.3 probably provides a good benchmark from which to assess the prospects for future supply – which themselves will be heavily affected by the conditions of future demand.

[18] Between 1970 and 1990 per capita cereal consumption in the Far East rose by about 90 kg. Total (i.e. direct and indirect) per capita cereal consumption in North Korea around 1990 was about 490 kg. It might be objected that Japan's level of cereal consumption per head (around 290 kg) is much below all these figures. But if account is taken of the fact that Japan (a) imports significant quantities of meat and (b) consumes a disproportionately large share of the world's fish catch, then its equivalent cereal consumption may actually be as high as 550 kg per head (see Gilland 1993, p. 87). In 1992 only about one-fifth of China's cereal harvest was fed to livestock – indicating considerable scope for increased indirect grain consumption.

[19] Of course, the point is partly that the composition effect – of faster demographic growth in lower per capita consumption regions – has to be overcome. This effect also contributed to the recent decline in world per capita cereal production examined in chapter 3.

FUTURE SUPPLY OF CEREALS

We start this consideration of the world's ability to expand its food production to meet this growing demand with a brief examination of the basic physical factors of agricultural production – namely the supply of cropland, the extent of irrigation and the level of fertilizer use (see Table 4.4).[20]

Basic agricultural resources: cropland, irrigation, fertilizers

The scope for meeting future rises in food demand simply through an expansion of the area of cropland is rather limited. Table 4.4 compares FAO estimates of areas of 'potential' arable land with the areas currently cropped.[21] It shows that about 45 per cent of the world's present cropland lies in Europe/FSU and North America/ Oceania – a fact which helps explain the high levels of per capita food production of these regions. In the Middle East and South Asia virtually all of the estimated potential arable land is already being cropped. And the same is true throughout most of the Far East, where the major reserves of unexploited land are chiefly found in sparsely populated parts of Indonesia (e.g. Kalimantan) and Indochina (e.g. Cambodia).

The circumstances of sub-Saharan Africa and Latin America are somewhat different. In these regions the statistics suggest that under one-fifth of potential arable land is currently being cropped (Table 4.4), although because of data problems the true fractions may be somewhat higher.[22] However, the main tracts of unexploited land are in locations like the Congo and Amazon basins – which are quite far from the major regional population concentrations[23] – and

[20] All these factors receive further attention in chapter 5.
[21] The 'potential' areas were derived from an FAO Agro-Ecological Zone Study (see Alexandratos 1988, p. 327). The areas of cropland refer to land under both temporary and permanent (e.g. fruit trees, coffee) crops.
[22] For example, the cropland statistics in Table 4.4 should include all land that has been fallow for fewer than five years. But according to Grigg (1993, p. 75), in recent years there has been an increasing tendency for some African governments to exclude such land from their statistical returns.
[23] The main population concentrations in sub-Saharan Africa – with large agricultural areas often containing over 100, and sometimes over 250, people per square kilometre – are in Nigeria, the Ethiopian highlands, and the East African highlands (including Rwanda, Burundi, Uganda, and parts of both northern Tanzania and eastern Kenya). Similarly, a high proportion of Latin America's total population lives in the eastern coastal belt stretching between Recife (in Brazil) and Mar del Plata (in Argentina); this region's second major population concentration is in central Mexico.

Table 4.4 Factors affecting the future supply of cereals, world regions, around 1990

Region	*Potential arable land (million hectares)*	*Cropland, 1991 (million hectares)*	*% of potential arable that is cropped*	*% of cropland that is irrigated, 1989–91*	*Fertilizer use, 1989–91 (kg per hectare of cropland)*
Sub-Saharan Africa	815.7	142.0	17.4	2.9	13.9
Middle East	72.3	70.2	97.1	16.7	53.9
South Asia	230.2	227.6	98.9	33.2	73.2
Far East	235.9	185.2	77.9	38.1	196.8
Latin America	889.6	153.3	17.2	10.9	52.3
Europe/FSU	n.a.	366.9	—	10.1	131.5
North America/Oceania	n.a.	282.2	—	7.7	79.2
World	—	1441.6	—	16.8	98.5

Notes: (a) For the Middle East the above statistics on arable land and cropland exclude Sudan. This country alone contains about half the region's total estimated supply of 'potential' arable land, but only about 20 per cent of it is cropped. If Sudan is included, then only 61 per cent of the Middle East's potential arable land is cropped. Similarly, the corresponding statistics for the Far East exclude Mongolia. However, the irrigation and fertilizer statistics shown above are inclusive of Sudan and Mongolia. Because estimates of China's potential arable land were unavailable, for convenience it was assumed that all such land there is cropped

(b) The above statistics on cropland and irrigation refer to net areas, i.e. irrespective of whether or not they are multi-cropped

(c) The irrigation statistics refer to areas purposely provided with water, including areas irrigated by deliberate flooding. Fertilizer use refers to the application of nitrogen (N), phosphate (P_20_5) and potash (K_20). Of these N is by far the most important; in developing countries it is mainly applied as solid urea ($CO(NH_2)_2$)

Principal sources: FAO estimates of potential arable land are from Alexandratos (1988); remaining estimates are chiefly devised from World Resources Institute (1994)

their systematic exploitation for agricultural purposes may therefore be restricted by considerations of distance and poor communications. Also, when such tropical forests are cleared for food production their soils often turn out to be not very productive over the long run. This said, as in the past, so in the future, it seems highly likely that, particularly in sub-Saharan Africa, a significant fraction of the increasing demand for food will be met by expanding the area of cropland. This process will probably involve continual expansion outwards from the main existing rural population concentrations, plus some long-distance migration to more remote and less exploited areas.

However, looking at the world as a whole, it is clear that the vast majority of increased food demand will not be met by the simple conversion of new land to crops. In any case much of the unexploited land which exists is probably of lower agricultural potential than that being lost to factors like urban growth.[24] Instead, the overwhelming bulk of future food demand must be met through more productive use of existing cropland.

Turning to irrigation, Table 4.4 shows that by 1990 about 17 per cent of the world's total cropland was irrigated. This represents some 242 million hectares – more than double the area that was irrigated in the 1950s. At the global level, average crop yields on irrigated land are roughly three times greater than yields on unirrigated land, and it has been estimated that 36 per cent of the current world harvest comes from the irrigated area.[25]

The Far East and South Asia have by far the largest fractions of cropland under irrigation (see Table 4.4). Indeed, China and India have respectively 48 million and 46 million hectares of irrigated land – from which both countries produce more than half their total domestic output of food. Other major irrigators are the former Soviet Union (21 million ha), the United States (19 million ha), Pakistan (17 million ha), Iran (6 million ha) and Mexico (5 million ha).[26]

However, it is generally agreed that from either an engineering or

[24] Some of the world's major fast-growing cities (e.g. Cairo, Mexico City, Beijing) owe their very location to the agricultural productivity of the soils on which they sit.

[25] See Postel (1992, p. 49). On the relationship between irrigation and rice yields in Asia see Grigg (1993, pp. 219–24).

[26] These country estimates refer to 1991 and relate to the net irrigated area, i.e. they take no account of repeat irrigation due to multiple cropping. See FAO (1993, p. 16).

economic perspective, many of the best opportunities to develop large and medium-scale irrigation systems have been used up. Therefore some analysts believe that the future expansion of the world's irrigated area will be slower than in the past.[27] For example, it has been reported that China and India both plan to exploit virtually all of their remaining irrigable capacity by the year 2000.[28] Likewise, in the Middle East there are few remaining reserves of land with easy irrigable potential, and water is often a matter of dispute between states. In sub-Saharan Africa irrigation is found in only a few countries (e.g. South Africa, Madagascar and Nigeria) and is often used for export crops rather than for domestic food production. Large areas of Sahelian, eastern and southern Africa already face severe water scarcity – and not just for agriculture.[29] Moreover, competition for water resources from fast-growing towns is widespread, especially in the developing world. So, again, the conclusion must be that most of the future increase in food demand will have to be met through greater efficiency of use of current water supplies.

This brings us to the statistics on fertilizers, of which by far the most important are nitrogenous. Roughly half of all nutrients incorporated into the current world harvest come from the application of synthetic nitrogen fertilizers. And provided there is sufficient water, nitrogen availability is the single most important determinant of crop yields.[30] Regional differences in fertilizer use broadly correspond to differences in agricultural intensification (see Table 4.4). Fertilizer use is especially high in the Far East. Between 1980 and 1990 China doubled its average applications from around 144 kg to 284 kg per hectare of cropland – so helping to fuel its sharp rise in food output. Indeed, of the 67 developing

[27] See, for example, Crosson and Anderson (1992, pp. 45–54) and Postel (1992, 1994). This issue is treated further in chapter 5.
[28] See World Bank and United Nations Development Programme (1990) *Irrigation and Drainage Research: A Proposal*, Agriculture and Rural Development Department, World Bank, Washington, DC, cited in Crosson and Anderson (1992, p. 46 and p. 122). However, this reported statement should be regarded with caution since it is unclear as to the type of irrigation capacity it relates. Thus Postel (1992, p. 55) reports that India's entire irrigation potential is 113 million hectares, an area so large that it will certainly not be developed by 2000. As so often, statements and statistics on agricultural matters must be interpreted with great care.
[29] See Falkenmark (1991).
[30] See Gilland (1993) and Smil (1991, p. 577). Around 1990 synthetic nitrogen comprised about 55 per cent of world fertilizer production by weight. All fertilizer statistics are taken from World Resources Institute (1994).

countries in our sample only Egypt (360 kg), North Korea (410 kg) and South Korea (440 kg) used more fertilizer per hectare around 1990 than did China. Average levels of application are also relatively high in Europe/FSU, reaching maxima in Germany (520 kg) and the Netherlands (614 kg).

Although the international price of urea – the principal nitrogen fertilizer in most developing countries – has fallen considerably in recent years, the energy and raw material costs for its production are still quite high.[31] This helps explain why levels of fertilizer use are generally much lower in the other developing world regions (Table 4.4). Also, as we have already noted, the yield gains to be had from synthetic fertilizers are greatest where there is irrigation. Conversely, in conditions of high temperatures, porous soils or heavy rainfall (such as apply, for example, in parts of sub-Saharan Africa) the benefits of synthetic fertilizers can be reduced as they are either oxidized or leached away.

Regional differentials in cereal yields reflect a host of factors – including variation in land quality. Also, many considerations – such as the type of fertilizer used, and the placement and timing of application – influence the fraction of synthetic fertilizer nutrients which are actually absorbed by the plants. Nevertheless, the strength of the basic relationship between fertilizer inputs and cereal yields shown in Figure 4.3 is unmistakable. It provides powerful support for the view that, provided collateral inputs are forthcoming, there exists considerable scope to raise cereal yields through greater fertilizer use – especially in countries in sub-Saharan Africa, the Middle East, South Asia and Latin America.

The relative contributions of area and yield

If demographic growth is set to be the dominant determinant of increasing world cereal demand, it should be equally clear that yield growth will be vital to the future expansion of world cereal supply. We have already seen that rising yields have accounted for most of the increase in global cereal output during recent decades. Table 4.5 shows that this was true for every world region.

[31] Urea is often favoured because as a solid it is easier to transport, store and apply than is ammonia gas (ammonia is widely used in maize cultivation in the United States and helps to account for that country's exceptionally efficient use of nitrogen in agriculture). The world price of urea in constant 1990 $US fell from $309 in 1980 to $132 in 1992. See World Resources Institute (1994, p. 262).

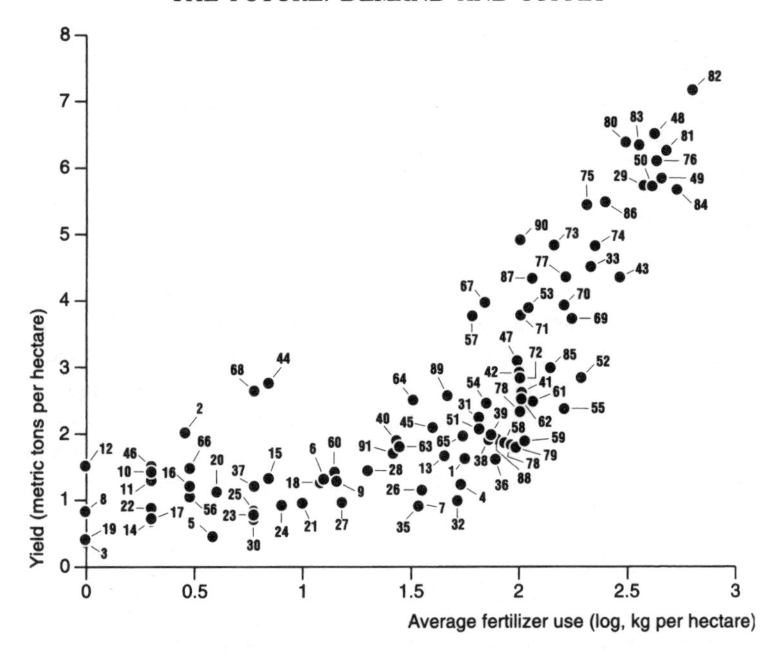

Figure 4.3 Fertilizer use and cereal yields, sample countries, 1989–92
Note: Country numbers are given in Figure 2.1 and in the Appendix

However, these statistics on the contributions of area and yield to past cereal production growth must be interpreted with care. Just because certain regions have few reserves of potential arable land which can be converted into crops does not preclude some extension of their harvested area through greater multiple cropping. Also, although the 1980s saw some sharp reductions in harvested area – especially in North America/Oceania and Europe/FSU – nevertheless, most regions still experienced positive growth rates of area harvested of cereals over the full period 1970–90 (see Table 4.5).[32]

Cereal production trends during the 1980s in the two most economically developed regions have, of course, been heavily influenced by reductions in harvested area – which inevitably

[32] The negative growth rate shown in Table 4.5 for Europe/FSU is mainly due to reductions of harvested area in the former Soviet Union – partly reflecting increased alternate-year fallowing there.

Table 4.5 Percentage contributions of area and yield to total cereal production growth, and growth rates of harvested cereal area, world regions, 1970–90

	1970–80		*1980–90*		*1970–90*		*Average annual rate of change (%) of harvested cereal area, 1970–90*
	Area	*Yield*	*Area*	*Yield*	*Area*	*Yield*	
Sub-Saharan Africa	23	77	26	74	23	77	0.54
Middle East	33	67	40	60	34	66	1.00
South Asia	19	81	0	100	7	93	0.26
Far East	8	92	1	99	4	96	0.17
Latin America	21	79	−32	132	6	94	0.15
Europe/FSU	18	82	−70	170	−27	127	−0.43
North America/ Oceania	51	49	0	100	24	76	0.49
World	22	78	−16	116	6	94	0.15

Notes: (a) All data used were based on three year averages, e.g. 1970 = 1969–71
(b) For the method employed in calculating these percentage contributions see the notes to Table 3.1. During the period 1980–90 in North America/Oceania there was only a small increase in cereal production (1.6 million tons) but a massive decline in harvested area (−9.7 million hectares) which, given the method employed, leads to 2167 per cent of the production increase being attributable to yields and −2067 per cent being attributable to area. For simplicity, and to avoid this anomaly, above we have simply assigned all of the production increase to yield growth

raises the proportional contributions which are attributable to yields (Table 4.5). But in both these regions there are significant reserves of land which could be returned to cereal cultivation in the future if required.[33] Similarly, Latin America's decline in harvested area between 1980 and 1990 was entirely due to Argentina. In the rest of Latin America there was actually a slight rise in the area harvested of cereals, contributing some 10 per cent to the overall regional increase in cereal production which occurred outside of Argentina.[34]

Despite the indications of Table 4.5 it is probable that in sub-

[33] In 1993 the United States held at least 8 million hectares of land out of cereal production under its commodity supply programmes alone. See Brown (1994a, p. 182).

[34] Likewise, if Argentina is excluded from the calculations for 1970–80 in Table 4.5 then in the rest of Latin America area expansion accounted for 29 per cent of cereal production growth during that decade.

Saharan Africa yields have made only a modest contribution to total cereal output growth. Although the region's harvested area seems to have grown at a comparatively fast rate over the period 1970–90, there are reasons to believe that the available statistics understate the true rate of area expansion. In turn, this probably inflates the apparent contribution made by yields. Certainly, many authorities believe that area expansion – rather than yield growth – accounts for most of sub-Saharan Africa's past gains in cereal output.[35]

Finally, even in the most land-scarce world regions some expansion of harvested area has occurred during recent decades. The extent of this enlargement is perhaps most surprising in the Middle East, apparently averaging 1.0 per cent per year during 1970–90 (see Table 4.5). In the 1980s several countries in this region (e.g. Syria, Morocco, Sudan, Iraq, Saudi Arabia and Egypt) considerably expanded their cereal area. They did this partly through increased multiple cropping, and by reclaiming and irrigating significant tracts of steppe and desert land. Similarly, in the mid-1990s Turkey's opening of the Southeast Anatolia Project will massively enlarge that country's area under irrigation. However, even in South Asia and the Far East – where yields accounted for virtually all cereal production growth during 1980–90 – there was still some minor area expansion over the longer run (Table 4.5). In the future, even these regions may marginally augment their cereal area, for example through increased multiple cropping.[36]

So while the key to future world cereal production certainly lies in higher yields, the likelihood of some minor contribution from area expansion – somewhat obscured by the particular experience of the 1980s – should also be borne in mind.

Yield levels and trends

It is obvious that yields constitute the really critical issue. In this context Table 4.6 presents the data on yield levels and trends which will underpin our calculations regarding future cereal production.

Two related conclusions from chapter 3 were (i) that aggregate

[35] See Grigg (1993, p. 88).

[36] Postel (1989, p. 30) refers to estimates that in India irrigation management improvements could enable an additional 8 million hectares of land to be irrigated. This would represent almost a 20 per cent increase in the country's irrigation capacity without developing any new water sources.

Table 4.6 Average cereal yields, levels, trends and projections, world regions, 1951–2020

Region	*Average yield around 1990 (mt per ha)*	*Average annual cereal yield increase (kg per ha per year)*				*Projected yield in 2020 based on 1981–93 increment (mt per ha)*
		1951–93	*1961–93*	*1971–93*	*1981–93*	
Sub-Saharan Africa	1.165	10.2	10.8	11.7	14.4	1.597
Middle East	1.642	19.1	21.0	23.4	31.3	2.581
South Asia	1.919	27.4	32.9	40.4	50.6	3.437
Far East	3.817	67.0	82.7	92.3	72.7	5.998
Latin America	2.119	30.2	34.2	38.9	32.7	3.100
Europe/FSU	2.816	45.7	47.4	43.3	47.7 (59.1)	4.247
North America/Oceania	3.734	55.5	50.0	42.5	42.4	5.006
World	2.711	40.9	45.2	45.5	42.6	3.989

Notes: (a) 1990 yields are averages for sample countries for years 1989–91

(b) Average annual yield increases are the slopes of the corresponding linear regression lines for each period. Overlapping rather than decadal periods have been used here to lend a measure of stability to the comparisons; but this does not materially affect the conclusions drawn in the text

(c) The 1981–93 average cereal yield increase for Europe/FSU is given in parentheses. This figure of 59.1 kg is biased upwards by an exceptionally high regional yield recorded for 1993. This was largely due to a reduction in the area planted in the former Soviet Union which means that Europe (where yields are higher) is overrepresented in the 1993 regional composition. Therefore for present purposes the lower yield increment of 47.7 kg for the period 1981–92 has been preferred

yield changes are probably better depicted as a linear function of time, and (ii) that there is no reason to be especially anxious about recent yield trends. Both these conclusions are broadly supported by Table 4.6.

At the global level the average cereal yield increment since 1981 has been about 42 kilograms per hectare per year – a figure which is close to the long-run average increment since 1951 (see Table 4.6).

The same comparison at the regional level shows that those regions with the lowest yields – sub-Saharan Africa, the Middle East and South Asia – have each experienced rises in yield increments during more recent time periods. True, the average annual increment for sub-Saharan Africa since 1981 (14 kg) is pitifully small. But the corresponding figure for South Asia (51 kg) is fairly respectable in comparative terms and is much greater than in any previous time period (see Table 4.6).

Average annual yield rises in the Far East have evidently slackened somewhat in the period since 1981 and they have not increased in Latin America. Nevertheless, these rises remain slightly higher than the long-run average gains since 1951. The same applies for Europe/FSU even if, as is probably preferable, we use the average increment (48 kg) for the period 1981–92.[37]

North America/Oceania is the only world region where there is any sign of a yield slowdown compared to the long-run average increment since 1951. In this region yields have been increasing at about 42 kg per hectare per year since about 1971.

Table 4.6 also shows what cereal yields will be in the year 2020 if the average increments of the period 1981–93 continue into the future. (Recall that it is yield growth in precisely this period which has generated such anxiety.) Linear extrapolation suggests that the average world cereal yield in 2020 will be roughly 4 tons per hectare. Regional yields will vary from under 2 tons in sub-Saharan Africa to almost 6 tons in the case of the Far East (see Table 4.6).

The issue, therefore, is whether these projected yields – derived on the basis of the period 1981–93 – are sufficient to match the volume of projected cereal demand in the year 2020. We need to weigh considerations of both production and consumption, yield and area, and take account of potential inter-regional cereal transfers (i.e. trade and aid). This is the task of the next section.

[37] See the notes to Table 4.6 for explanation of this preference.

WEIGHING 'DEMAND' AND 'SUPPLY'

Building upon the foregoing discussion of future demand and supply, Table 4.7 attempts to provide a plausible scenario which broadly reconciles all the main quantities that are involved. The scenario hinges around the combination of different assumptions regarding future changes in yields and harvested area. We begin from an initial presumption that in most world regions the area harvested of cereals in the year 2020 will be the same as was harvested around 1990. We will consider the implications of Table 4.7, each region in turn.

The importing regions

Whether measured in percentage or linear terms both sub-Saharan Africa and the Middle East have experienced some acceleration of yield growth in the period since the early 1980s (see Tables 3.2 and 4.6). But their performance is still poor. And we can be fairly sure that neither region can meet its future cereal demand growth from its own production capacity.

Dealing first with sub-Saharan Africa, Table 4.7 gives the projected 2020 regional cereal demand of about 165 million tons. Recall that this figure is based solely upon population increase. We have seen that in this region it makes little sense to consider increases in cereal production without taking both yield growth and area expansion into account (partly because the data confound the two influences). Accordingly, in Table 4.7 it has been assumed that sub-Saharan Africa's harvested area will continue to expand at the same average annual rate as during 1970–90 (see Table 4.5) – in which case it will reach around 69.8 million hectares by the year 2020. With a projected average cereal yield of 1.597 tons this would produce about 112 million tons of cereals. And if we assume an average yield increment of 20 kg per hectare per year – certainly optimistic, but perhaps not impossible given the small rises in past increments conveyed by Table 4.6 – the average yield in 2020 would still be only about 1.765 tons, with corresponding total regional production of 123 million tons of cereals. These figures suggest that by the year 2020 sub-Saharan Africa may need to import in the vicinity of 25 to 32 per cent of its total cereal consumption. In turn, this means that net transfers of grain into

Table 4.7 A plausible 2020 scenario: reconciling projected regional cereal demand and supply

Region	Projected demand (million mt)	Area harvested (million ha)	Projected 2020 Yield (mt per ha)	Projected 2020 Production (million mt)	% of consumption/demand met by imports and aid 1990–92	% of consumption/demand met by imports and aid 2020	Net cereal volumes imported by trade and aid (million mt) 1990–92	Net cereal volumes imported by trade and aid (million mt) 2020
Sub-Saharan Africa	164.8	59.3 (69.8)	1.597 to (1.765)	111.5 to 123.2	9.2	25 to 32	6.9	42 to 53
Middle East	218.4	40.3 (46.8)	2.581	120.7	30.6	45	32.2	98
South Asia	532.6	140.3	3.437	482.4	4.3	9	12.1	50
Far East	1004.9 to (1128.8)	145.1	5.998	870.0	8.4	13 to 23	50.9	135 to 259
Latin America	208.1	48.4 (50.6)	3.100	150.0 to 177.1	10.0	15 to 28	11.7	31 to 58
Europe/FSU	521.6	171.4	4.247	727.7	2.3	−39	11.5	−206
North America/ Oceania	309.3	98.4	5.006	492.7	−54.5	−59	−128.9	−183
World	2959.7	722.4	4.091	2955.0	—	—	—	—

Notes: (a) Projected demand is taken from Table 4.3. If the higher demand figure for the Far East is used then total world cereal demand in 2020 would be about 5 per cent higher at 3083.6 million mt

(b) Except for the figures in parentheses, the regional areas harvested shown are averages for years 1989–91. They differ slightly from the estimates shown in Table 3.2 due to differences in period and because they include allowance for non-sample countries. The figures in parentheses assume some expansion of the harvested cereal area and have been used in calculating the world total; see the text for their derivation

(c) Projected 2020 regional yields are taken from Table 4.6. See the text for the derivation of the figure in parentheses for sub-Saharan Africa

(d) Total cereal production in 2020 is based on the lower production estimates for sub-Saharan Africa and Latin America. The projected average world cereal yield for 2020 is based on the area and production figures shown above and therefore differs slightly from the estimate shown in Table 4.6

(e) Percentage of consumption met by trade and aid in 1990–92 is taken from Table 3.4. Combined with the regional consumption estimates in Table 4.3 these proportions give the import volumes shown above. Negative numbers denote net exports and donations as a proportion of consumption; these percentages would be smaller if they were expressed in relation to production

(f) All numbers should be regarded as only very broadly indicative. There is no intention of detailed consistency

the region will need to rise from nearly 7 million tons in 1990–92 to around 42 million to 53 million tons (see Table 4.7).

The food prospects facing sub-Saharan Africa are sometimes characterized in what seem overly alarming ways.[38] But however the numbers are manipulated, the region appears incapable of increasing its food output to match its demographic growth. It seems inevitable that sub-Saharan Africa confronts a future of increasing dependence upon food imports and aid, and either roughly constant or even declining levels of per capita consumption too.

We have projected cereal demand in the Middle East at around 218 million tons, and we have seen that the region's recent yield growth has also improved. With greater fertilizer and other inputs it is certainly feasible to expect that average yields could approach 2.581 tons per hectare by the year 2020 (see Tables 4.4 and 4.6). What looks much more problematic is that the harvested area can continue to expand at the seemingly remarkably high average annual rate achieved during the period 1970–90. Such an assumption seems unlikely partly because in the absence of another oil boom the investment resources for such an expansion (largely in irrigation) will not exist, and, still more, because of the region's rapidly tightening general water constraints. Accordingly, we have arbitrarily assumed that the rate of expansion of harvested area will be 0.5 per cent per year (i.e. half the previous compound rate) – still a high figure in comparative terms (see Table 4.5). In turn, this implies a harvested cereal area of 46.8 million hectares in the year 2020, with a corresponding regional cereal output of around 121 million tons. Whatever may be the validity of these particular assumptions, it seems highly probable that by the year 2020 the Middle East will depend upon outside sources for nearly half of its

[38] For example, referring to a publication by the World Bank (1993b) Pinstrup-Andersen writes that: 'If current trends continue, then the World Bank estimates that by the year 2020, [sub-Saharan] Africa will have a food shortage of 250 million tons, which is more than 20 times the [region's] current food gap.' See Pinstrup-Andersen (1994a, p. 21; also 1994b). 250 million tons (maize equivalent) is more than all cereals currently traded internationally. However, it is important to stress that the World Bank's estimates of food shortage are based upon assumed levels of general food requirements, and, more importantly, that the World Bank also provides alternative estimates of the 2020 food gap which are respectively 110 and only 5 million tons. See World Bank (1993b, pp. 73–5). So the 'estimate' of 250 million tons mentioned by Pinstrup-Andersen is actually only one of several illustrative alternatives to which, unfortunately, he does not refer.

total cereal supplies – perhaps exceeding 90 million tons of grain each year (see Table 4.7).

The results for South Asia are rather more promising. Table 4.6 shows that were the recent regional yield increment of about 51 kg per hectare per year to continue, then the average yield in the year 2020 would be 3.437 tons. Assuming no change in the area harvested, then the corresponding volume of regional cereal production would be about 482 million tons – leaving about 9 per cent of demand to be met by transfers. Although such a scenario still involves an annual inflow of some 50 million tons of cereals (probably mostly going to Iran, and perhaps Bangladesh),[39] the basic numbers for South Asia look comparatively encouraging (see Table 4.7). This is partly because the region's demographic growth is generally slowing, and partly because there is certainly considerable scope for greater fertilizer use to raise yields significantly.[40] Moreover, we should stress that perhaps our assumptions regarding future production prospects have been conservative in that they involve no area expansion and a yield increment which could well rise further in the coming decades (see Tables 4.6 and 4.7).

The prospects for the Far East are more difficult to fathom, partly because of the greater margin of uncertainty regarding the volume of future demand. Table 4.6 shows that if yields continue to increase at the average increment experienced since 1981 they will reach 6 tons per hectare by 2020. Although this figure is certainly high for a regional average, it could well prove feasible over the thirty-year period being considered here. It is close to the yield level already reached by North Korea, South Korea, Japan and several major European countries, including France.[41] Assuming no change in harvested area, this yield would produce an annual cereal harvest of about 870 million tons by the year 2020 (see Table 4.7).

[39] See chapter 6 and Dyson (1993).

[40] 'There are still some countries, such as India . . . where there is a large potential for profitably boosting fertilizer use.' See Brown (1991, p. 13). See also Duddin and Hendrie (1992, p. 97).

[41] Average cereal yields for North Korea, South Korea, Japan and France in 1990–92 were respectively 6.497, 5.808, 5.704 and 6.372 tons per hectare; the corresponding fertilizer inputs were all in excess of 300 kg per hectare. See Plucknett (1993) and World Resources Institute (1994). It is worth noting that Crosson and Anderson's (1992) projections may imply an average world rice yield of about 6.8 tons in the year 2030, though the plausibility of such a high figure has been questioned. See Gilland (1993, p. 87).

Of course, the adequacy of a cereal harvest of this size is dependent upon which projection of demand we accept. The figures in Table 4.7 imply that the Far East may have an annual import requirement of anywhere between 135 and 259 million tons. These figures are equivalent to 13 and 23 per cent of projected demand respectively. For reasons which should already be apparent, we give precedence to the lower range of these figures. This said, it seems inevitable that the Far East will greatly expand its volume of cereal imports in the next few decades. And it seems inevitable that the region will become significantly more dependent upon imports as a fraction of its total cereal consumption.

Finally, among the likely future cereal importing regions, we come to Latin America. We have already seen that this region's recent experience has been heavily influenced by the particular circumstances of Argentina, which has had much of its own market share within Latin America undercut by the two big North American producers. Also, cereals are less representative of the general food situation of this region than is generally the case elsewhere. Nevertheless, Table 4.6 shows that if Latin America's average yield increment since 1981 of about 33 kg per hectare per year continues, then by 2020 the regional cereal yield will be around 3.1 tons per hectare. Assuming no change in harvested area this would produce around 150 million tons of cereals, leaving an import requirement of about 58 million tons – equivalent to 28 per cent of annual consumption.

Clearly, as in all regions, what exactly happens will be partly conditioned by trends in international prices. But Latin America's production prospects probably have a degree of flexibility which simply does not exist in the other cereal importing regions. For example, we have noted that there are clear possibilities to expand the region's harvested area. Thus if the harvested area were to continue to grow at the rather slow rate of 0.15 per cent per year experienced between 1970 and 1990 (see Table 4.5) it would reach 50.6 million hectares by the year 2020 (though the full potential for area expansion in Latin America is certainly very much larger still).[42]

[42] This statement can be supported, for example, by the considerable recent area reduction in Argentina and that country's obvious capacity to greatly increase its harvested area of wheat if international prices rise. There is also great scope for increased cereal output in Brazil – one significant component of which is the considerable potential for an expansion of rice cultivation in the Pará and Maranháo regions.

Likewise, especially with greater fertilizer use, an average regional cereal yield approaching 3.5 tons per hectare could well be attained. When combined, these admittedly rather arbitrary figures imply a 2020 cereal harvest of approximately 177 million tons, and an annual import requirement of 31 million tons – equivalent to about 15 per cent of demand (see Table 4.7). But it is not inconceivable that this percentage could be much smaller still if world cereal prices were to rise sufficiently.

The exporting regions

The foregoing scenario for the five cereal importing regions sketched above – and summarized in Table 4.7 – strongly suggests that they can all expect to meet a rising proportion of their future food consumption through imports. Indeed, in most regions this proportion could easily double over the time-horizon of concern here. Moreover, whereas around 1990 these regions collectively imported roughly 120 million tons of cereals from outside, Table 4.7 suggests that this annual demand requirement is set to greatly increase.

By exactly how much this collective 'demand' for imports will rise is hard to say; as we stated at the beginning of this chapter, demand and supply will inevitably interact. It could be argued that we have underestimated future yield growth. And much will depend upon demand developments in the Far East. However, there must be a good chance that by the year 2020 these five regions together will be importing roughly 375 million tons of cereals each year, although the figure could be somewhat more or less.[43]

The key question, therefore, is whether the two remaining world regions – Europe/FSU and North America/Oceania – can possibly meet these greatly heightened export requirements.

On the basis of past trends and experience, the answer must surely be 'Yes'.

Table 4.7 shows that if the annual yield increments experienced since 1981 continue into the future then average cereal yields in Europe/FSU and North America/Oceania will be about 4.247

[43] For simplicity, this figure of 375 million tons is based on the average of the two cereal import volumes shown for sub-Saharan Africa and Latin America in Table 4.7, plus the preferred lower figure shown for the Far East.

tons and 5.006 tons per hectare respectively. The corresponding total production figures are roughly 728 million and 493 million tons. When these production estimates are compared with the projected demand figures they imply respective cereal 'surpluses' of 206 million and 183 million tons. The combined surplus of about 389 million tons is sufficient to meet the envisaged collective requirements of the five importing regions.

Of course, all these figures convey an entirely unwarranted impression of solidity and precision! But they do broadly imply that, together, Europe/FSU and North America/Oceania probably *do* have the capacity to meet a plausible expansion in global food demand during the next two or three decades. While it is true that the growth in demand might exceed the figure of 375 million tons which we have entertained here, it is also reasonable to suggest that we have understated the capacity of the two exporting regions to raise their yields (and therefore output) in the years ahead. This possibility may be strongest in the case of North America/Oceania. Certainly, both regions also have scope to expand their cereal area.

Whether Europe/FSU will actually displace North America/Oceania as the world's leading cereal exporting region (as is implied in Table 4.7) is certainly very questionable. To some extent the issue will depend upon developments and yield trends in eastern Europe – where there are encouraging signs – and the former Soviet Union – which in the early 1990s contained about 60 per cent of Europe/FSU's total area harvested of cereals, but also had the lowest yields.[44] However, in general there is little doubt that the attainment of average yields of between 4 and 5 tons per hectare in the year 2020 is a realistic expectation for both regions. Indeed, in the US and much of Europe average cereal yields in the early 1990s were already well within this range, and rising.[45] Moreover, it should be recalled that all these calculations have assumed a constant cereal area – although there is certainly room for some expansion through reduced set aside. But should actual yield growth be faster, and/or per capita cereal consumption in the

[44] The average cereal yield in the territory of the former Soviet Union in 1990–92 was 1.778 tons per hectare. The next lowest yield among the region's sample countries was in Portugal, at 1.862 tons.

[45] The average yield in the US in 1990–92 was 4.874 tons per hectare. Excluding eastern Europe and the FSU, the figure for the rest of Europe was 4.738 tons. Both these figures will certainly increase.

developed world decline (e.g. due to changes in meat intake), then the demand requirements of the importing regions might be met even with some reduction in the harvested area.

OVERVIEW OF THE SCENARIO

Table 4.7 strongly suggests that a future credible demand scenario can be matched by supply. That is, a realistic increase in the volume of cereal demand – mostly due to population growth, but also allowing for a non-demographic component – can be met on the basis of fairly defensible, perhaps even conservative assumptions as to future production trends. Recall that for the four most important cereal producing regions no expansion of harvested area has been assumed in these calculations. But even in the most land-scarce regions there clearly is scope for such expansion – for example through greater multiple cropping. And both Europe/FSU and North America/Oceania could expand their cereal areas, if required. Also, it may well be that our assumptions have understated the yield increments which can be expected for some importing regions.

The scenario envisages an average world cereal yield of about 4.0 tons per hectare by the year 2020 (see Tables 4.6 and 4.7). Since this figure is largely based on the continuation of an 'arithmetical' (i.e. linear) trend it is inevitable that annual percentage increases in the global yield will tend to decline in the years ahead. In fact, the present scenario implies that the world cereal yield will increase at approximately 1.3 to 1.4 per cent per year over the period 1990–2020, a compound growth rate which is not out of line with the range of other projection estimates.[46] With increased inputs, especially of fertilizers, there is nothing obviously far-fetched about the average yields envisaged in Table 4.7.

Of course, the scenario is not without its problems. The volume

[46] The range of '1.3 to 1.4' is given because small differences in the projected 2020 yield (as in Tables 4.6 and 4.7) translate into sizeable differences in average annual percentage growth rates (another reason to treat such growth rates with care). Crosson and Anderson (1992, p. 12) project that between 1988–89 and 2030 world grain consumption (with rice measured in hulled form) will rise from 1677 to 3297 million tons, i.e. at 1.65 per cent per year. Mitchell and Ingco (1993, p. 180) forecast that world grain output will rise at 1.29 per cent per year over 1990–2010. Gilland (1993, pp. 86–7) provides two possible figures for world cereal output in 2030 (3083 and 3280 million tons) which imply average annual growth rates of 1.14 and 1.29 per cent respectively.

of global cereal transfers seems set to massively increase. The importing regions will all become still more dependent upon transfers – in proportional as well as absolute terms. It seems inevitable that the Far East will become an even more dominant region in determining the overall volume of world cereal demand, followed by the Middle East. Unless the interests of sub-Saharan Africa are deliberately protected in some way, it appears likely that levels of per capita cereal consumption in the region could decline still further. On the other hand, suggestions that sub-Saharan Africa will experience a twenty-fold increase in its food gap by the year 2020 are implausibly dire.[47] And in some measure at least, Europe/FSU will likely emerge as a second major cereal exporting bloc.

To reiterate, the future will not unfold exactly to the tune of Table 4.7. The calculations are rough. The unexpected will occur. And the complexities and adaptabilities inherent in the relationships linking populations with their food supplies will inevitably come into play. Areas harvested, crops grown, yields attained, quantities of food imported, used to feed livestock, and directly consumed, will move in dynamic and often compensatory ways in different world regions. But the basic message is clear: issues of distribution apart, in the year 2020 the global agricultural system probably will be producing 'enough' food, at least in aggregate terms.

CONCLUDING REMARKS

The principal conclusions of this chapter can be stated briefly. In the period to the year 2020 the overwhelming majority of the increase in global cereal demand will be due to population growth. Therefore the future evolution of world food demand will be primarily a demographic (rather than an economic) matter. This is not to deny some continuation of the trend towards greater dietary diversity. But only in the Far East region, perhaps, will rising incomes have a major impact upon growth in overall food consumption.

Nevertheless, a general premise of this chapter has been that in all world regions levels of per capita cereal consumption will either remain constant or improve during the decades immediately ahead.

[47] See footnote 38.

Most of the evidence points in this direction. Sub-Saharan Africa seems to be the only region where average levels of food intake may actually decline – although this possibility may have been marginally exaggerated by our concentration upon cereals here. However, because those world regions with comparatively low levels of per capita food consumption will increase as a proportion of humanity (because their population growth is faster) significant, if modest, increases in regional per capita food intake are likely to be obscured in average statistics of *world* per capita consumption (and production). It is important to bear this fact in mind.

We have also seen that the average annual increment in world cereal yields since the early 1980s has been about 42 kg per hectare per year. This is actually slightly higher than the long-run average increment during 1951–93. There has been no obviously worrying yield 'slowdown'. There is every reason to expect that in all world regions average cereal yields will be markedly higher in 2020 than they are today; and a global average yield of around 4 metric tons in 2020 seems a perfectly realistic expectation. Increased use of fertilizers will certainly be a crucial enabling ingredient for this. Our present scenario anticipates that total world cereal production (and consumption) in 2020 will be in the vicinity of 3 billion metric tons. There may need to be some modest expansion of area harvested to meet this figure – but this will depend upon the detailed progress of yields.

However, it seems incontrovertible that there will also be an important regional mismatch between the evolution of future cereal demand and supply. The poorer cereal importing regions of sub-Saharan Africa, the Middle East, South Asia, the Far East, and perhaps even Latin America, look set to become increasingly dependent upon North America/Oceania, and Europe/FSU, for their basic food supplies. So the volume of inter-regional transfers – which are mostly traded – seems certain to hugely increase. Population growth will spur the future expansion of world food trade.

Of course, numbers – such as those we have considered here – are not everything. Undeniably they are rough-hewn. But the broad direction of their momentum, and their implications for developments during the next few decades, seem fairly clear. As we saw in chapter 1, the relationship between population and food is complex. It involves many flexibilities and adaptations. Yet, one way or another, the evidence is fairly persuasive that the demand growth which has been projected here for the year 2020 can, and probably will, be met.

However, we must look at the issue from other angles. This is done in chapter 5.

5

EXPLORING THE FUTURE: POTENTIALS AND CONSTRAINTS

INTRODUCTION

This chapter considers several sets of factors which will condition the world's food prospects in the period to 2020. Starting at the 'environmental' end of the spectrum, and working towards the more 'social' end, the subjects which are covered include global atmospheric changes, soil and water resources, farm inputs, crops, biotechnology and research, physical and human resources, and the nature of prevailing institutional structures. There is also discussion of possible trends in world food prices. Obviously, all these variables overlap and interrelate. And, equally obviously, they can be considered only in outline here.

Notwithstanding the apparent logic of the numbers we examined in the previous chapter, it is conceivable that there could be constraints which will prevent global food output rising in line with projected demand. Moreover, even if this is not the case, consideration of the overall context in which world food production must be raised should provide clues as to the plausibility of the scenario and general conclusions developed in chapter 4.

GLOBAL ATMOSPHERIC CHANGES

Greenhouse warming and stratospheric ozone depletion could have implications for future food production. For present purposes the first phenomenon is potentially much more important.

Global warming

According to the Intergovernmental Panel on Climate Change (IPCC) the average world surface air temperature has risen by

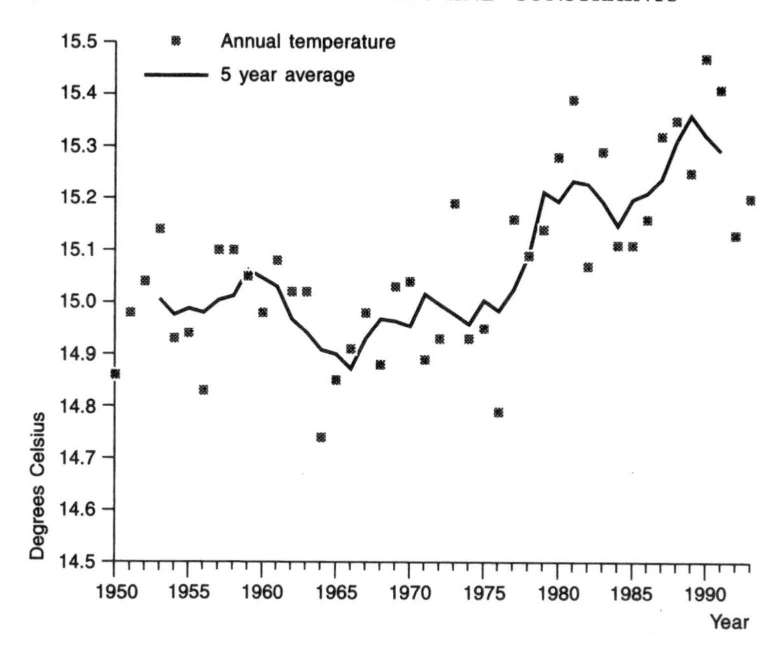

Figure 5.1 World average annual temperatures, 1950–93
Principal source: Roodman (1994a)

between about 0.3 and 0.6 degrees Celsius during the last 100 years.[1] The rise has been irregular – with comparatively rapid warming occurring between 1920 and 1940, and again since the mid-1970s. Starting in 1977 the average temperature has been above 15.00°C in every single year. 1995 has set a new high, but previously 1990 was the hottest year on record at 15.47°C (see Figure 5.1).[2]

There are strong grounds to believe that at least part of this warming is due to human-induced release of greenhouse gases (GHGs). Of these gases carbon dioxide ($C0_2$) – produced by the burning of fossil fuels, cement production and deforestation – is by far the most important. Next in importance come the chlorofluorocarbon gases (CFCs) – which are also the principal agents of ozone depletion (see pp. 142–3). Methane – which is emitted from

[1] This discussion draws on IPCC (1990a, 1990b, 1992, 1994). See also Hadley Centre (1993) and Henderson-Sellers (1994).
[2] For temperatures see Bloomfield (1992) and Roodman (1994a).

rice paddies and the guts of livestock – and nitrous oxide – which originates partly from nitrogen fertilizers and the creation of tropical grasslands – are also significant GHGs. Overall, agricultural activities may account for about one-sixth of total human-induced global warming.

However, there is ample uncertainty as to the extent of any future warming. There is even greater uncertainty about its likely nature – for example, its timing and regional impacts. There is still more uncertainty regarding its consequences for agriculture – for example, as to how rainfall patterns may change. And, really, there is only educated speculation as to how well agriculture may be able to adapt. Indeed, there is huge uncertainty not only about the future evolution of the ultimate causes of human-induced warming (e.g. economic growth, deforestation, energy use) but also about the crucial role which feedback mechanisms may play – whether they be physical (e.g. the role of clouds and oceans) or social (e.g. technological adaptations, or controlled reductions in the release of greenhouse gases).

The IPCC estimates that if nothing is done to control GHG emissions (the so-called 'Business-as-Usual' scenario) then the world's temperature could rise by between 0.6° and 1.5°C over the period 1990–2020. However, alternative scenarios which assume progressively increasing control of GHG emissions result in much lower average temperature rises, in the range 0.3° to 0.6°C. In this context it is worth noting that global production of CFCs – thought to be responsible for roughly a quarter of human-induced warming – has been sharply controlled since 1988.[3]

Moreover, all of these estimates of possible future warming are probably biased upwards because they make no allowance for other human-induced processes which may have a partially offsetting (i.e. cooling) effect. Most importantly, humanity is constantly stirring-up tiny particles into the atmosphere like smoke, dust and sulphur dioxide. These act as nuclei on which cloud droplets form. And this has a significant cooling influence by reflecting solar radiation back into space. In addition, stratospheric ozone depletion may itself have a minor cooling effect.[4]

[3] Following the Montreal Ozone Protocol of 1987 estimated world CFC production figures for 1988 and 1993 were respectively 1260 and 510 thousand tons, i.e. a 60 per cent reduction. See Ryan (1994, p. 64).

[4] See IPCC (1992, pp. 19–20; 1994, p. 24). Of course, increased particle emissions are linked to deleterious outcomes like acid rain.

As already remarked, the implications of any global warming for the conditions affecting world agriculture are largely a matter of informed guesswork.[5] It is generally agreed that some effects will be positive, and other effects will be negative. But the net balance is hard to judge.

Other things being equal, the growth of certain foodcrops may be faster because of the fertilizing effect of higher levels of carbon dioxide in the atmosphere. Experiments suggest that this effect could be quite pronounced for so-called C_3 crops like wheat, barley, rice and potatoes. But the effect would be much less for C_4 crops like maize, millet and sorghum. Moreover, an effect detected under experimental conditions may not manifest itself so readily in farmers' fields.[6]

Global warming could also threaten some of the world's cropland in river deltas and coastal areas – as the sea level rises due to the combined effect of thermal expansion of the oceans and melting glaciers. For example, significant parts of Bangladesh, Egypt, Thailand, China and Indonesia could eventually face this prospect, together with various complications such as increasing saltwater intrusion into freshwater supplies.

However, as is often the case regarding ominous prognostications about the future, there have been progressive downward revisions as to the magnitude of this anticipated threat. Whereas in the early 1980s it was considered possible that the sea might rise by several metres over the next 100 years, most recent estimates are in the range 20 to 80 centimetres.[7] The IPCC 'Business-as-Usual' scenario predicts a 'best estimate' average sea level rise of about 15 centimetres by the year 2020. But there could be significant, and largely unpredictable, regional variations to this figure – for example, because of quite unrelated movements in the vertical height of land.[8] Moreover, as we have seen, because it takes no account of possible control of GHG emissions, or compensatory cooling effects, it is probable that business will *not*

[5] For a summary and expansion of the IPCC assessment of the potential impact on world agriculture, see Parry (1990). See also Downing (1992) and Downing and Parry (1994).

[6] Plants are C_3 or C_4 according to whether they have a 3- or 4-carbon path for photosynthesis. Any fertilization effect is likely to diminish at levels of atmospheric CO_2 concentration above 600 parts per million. However, such levels lie far beyond the time-horizon being considered here.

[7] See World Bank (1992, p. 159).

[8] See IPCC (1990b, pp. 22–3).

continue quite as usual. Consequently, any sea level rise may be significantly less than 15 centimetres in the period of concern to us.

There is some consensus between the different climate simulation models that the temperature rises involved in global warming will be greater in the winter months – perhaps lengthening the growing periods for some crops – and that world rainfall will generally increase. Also, the temperature rises will probably be greater at higher latitudes – perhaps benefiting food production in northern areas of Europe, Asia and North America. On the other hand, higher temperatures could lead to reductions in soil moisture and soil fertility, and therefore increased risks of erosion. Such processes could reduce potential cereal yields in northern mid-latitudes like the US Corn Belt, southern Europe and parts of Ukraine and Kazakhstan in the former Soviet Union.[9]

It is particularly difficult to forecast the consequences of global warming for agriculture in the world's developing regions – which generally lie in mid and lower latitudes. There is little agreement between climate simulation models as to how levels and patterns of rainfall in these latitudes may be affected. It is possible that average annual rainfall in arid and semi-arid areas, like the African Sahel, could increase. But if this increase were to happen in just a few concentrated rainstorms, then there could also be greater frequency of flooding and little change in the risk of seasonal drought. Indeed, although it is generally agreed that global warming will probably involve an increase in average world precipitation, somewhat paradoxically the most serious consequence for global agriculture could well be a greater risk of drought.[10]

Finally, there is the issue of how well agriculture may adapt to any climate change. Here three preliminary points must be made. First, adaptation will obviously be easier in the face of *gradual* change – which, fingers crossed, is what is generally anticipated. Any sudden discontinuity would clearly be very much harder to

[9] See Parry (1990, p. 87).

[10] According to Parry (1990, p. 129) 'the potential impact of concurrent drought or heat stress in the major food-exporting regions of the world could be severe. In addition, relatively small decreases in rainfall, changes in rainfall distribution or increases in evapotranspiration could markedly increase the probability, intensity and duration of drought in currently drought-prone (and often food-deficient) regions'.

cope with.[11] Second, adaptation will inevitably involve costs. But provided change is gradual these are likely to be only a small component of the total global agricultural cost structure. Third, poor populations will generally find it harder to cope with any adaptation costs. This fact, coupled with the very tentative indications that populations at mid and lower latitudes may experience net negative effects from global warming, suggests that over the much longer run the adverse consequences for some developing countries may be significant.

Because, eventually, climate change could be so pervasive in its agricultural implications – affecting literally everything under the sun (e.g. soils, rainfall, storms, pests, weeds, livestock, etc.) – ultimately adaptations are likely to occur across the whole spectrum of farming activities. Other things being equal, food prices would rise because of increased costs. And the international trading system could be expected to function as a key adjustment mechanism, for example mediating if yields were to experience upward pressure in northern latitudes and downward pressure elsewhere.

Limited though they undoubtedly are, simulations which combine projections of the world's climate and agricultural production systems suggest that an eventual doubling of the level of atmospheric CO_2 (which is not expected to happen until well into the second half of the twenty-first century) may produce a small decline in world food production compared to what might happen in the absence of climate change.[12] On the other hand, if food prices were to rise sufficiently, stimulating greater investment in agriculture, then there could even be a modest positive effect on the total volume of world food output.[13]

[11] Of course, an abrupt rise in world temperature is not the only sudden calamitous possibility. Other, hopefully very slim, possibilities include a nuclear war, a meteorite hit or the rapid onset of an ice age. Apropos this last possibility Sugden and Hulton (1994, p. 150) write that 'We may yet discover that the rise in greenhouse gases associated with our use of the planet is the best way of preserving our equable present climate for the future.'

[12] For such a simulation see Rosenzweig and Parry (1994). They describe their results as 'relatively benign', although cereal yields in the developing world are adversely affected and international food prices rise. It should be stressed that such simulation results refer to a doubling of atmospheric CO_2 compared to concentrations prevailing in the pre-industrial era. Such a doubling lies well beyond the time-horizon being considered in the present book.

[13] See Kane *et al.* (1992). For an important analysis which suggests that global warming could have net economic benefits for US agriculture even without CO_2 fertilization, see Mendelsohn *et al.* (1994).

In summary, the effect of global warming on world agriculture is an uncertain and immensely complicated issue. But it seems unlikely that such warming will have much impact on world climate, and hence food and agricultural output, within the time-frame being considered by the present study. Any rise in global temperature before the year 2020 is likely to be gradual, and appreciably smaller than in the IPCC 'Business-as-Usual' scenario. The resulting adaptations of the world's agricultural production systems are likely to be largely subsumed within the continual process of agricultural development and change.

However, lest we be thought unconcerned, three words of caution are needed before we move on. First, to the extent that there are adverse consequences for food production before the year 2020 they may be felt most by the farmers of the developing regions. Second, if an increased risk of drought is indeed the most likely serious agricultural consequence of global climate change, then it is difficult to entirely banish from one's mind the recent sharp increases in harvest variability in North America/Oceania and sub-Saharan Africa which we documented in chapter 3 (pp. 87–9). Third, while the world may reasonably expect a comparatively smooth ride to the year 2020, the *eventual* longer run consequences of continued, and indeed rising, GHG emissions (especially CO_2) could be a lot more serious.[14]

Ozone depletion

The agricultural consequences of stratospheric ozone depletion will probably be much less significant than those of global warming.[15] Depletion of the ozone shield has chiefly occurred in mid and high latitudes. Therefore, to the extent that there are implica-

[14] CO_2 emissions involve an element of 'isolation paradox' in that while it may be in the world's interests to reduce emissions, it is not necessarily in the narrower interests of individual countries to restrict their emissions. During 1988–93 world carbon emissions have been approximately constant at about 5.9 million tons per year – reflecting economic recession and the break-up of the FSU. But in the mid-1990s the long-term upward trend will probably resume with economic recovery and fast economic growth in countries such as China. Yet a 60 per cent *reduction* in emissions is required just to stabilize atmospheric carbon levels. See Roodman (1994b, pp. 68–9).

[15] On this issue see World Bank (1992, pp. 62–3) and Ehrlich *et al.* (1993, pp. 17–18).

tions, they relate chiefly to agriculture in the world's more developed regions.

Because thinning of the ozone shield increases the ultraviolet-B (UV-B) radiation which reaches the earth's surface, it is possible that, were the process to continue, it might eventually affect agricultural production in northern areas of Europe, Asia and North America. There is certainly evidence of a seasonal reduction in vegetative phytoplankton in the Antarctic Ocean which, one day, may have harmful effects on some oceanic fisheries.[16] Perhaps more germane for our purposes are experiments which indicate adverse effects for the growth of certain crops. For example, soybeans are sensitive to UV-B radiation. But other plants, including rice, are fairly unaffected. And there is certainly scope for breeding food crops with enhanced UV-B resistance.

However, as already noted, global production of CFCs – the main agents of depletion – is now being sharply curtailed. Current predictions are that depletion of the ozone shield should peak by the year 2005, and that there will then be a period of slow recovery.[17] The chance that stratospheric ozone depletion will have a detectable and significant effect on aggregate levels of world food production before the year 2020 must be regarded as negligible.

SOIL AND WATER

Soil degradation and shortages of water for agriculture are both commonly mentioned as major impediments to increased output of food.

Soil degradation

Chapter 4 suggested that the world's harvested cereal area will probably expand by only a rather modest percentage during the

[16] The world catch of wild fish has recently been declining – but this has mainly been due to overfishing, plus the effects of coastal pollution, and less fishing by the (now) Russian fleet. This decline in capture fisheries has largely been offset by very rapid and continuing growth in world aquaculture production. Fish like tilapia and halibut seem set to form the basis of major industries. These topics are not treated in detail here. But note that aquaculture is broadly analogous to agriculture, in that output can usually be raised through greater allocation of resources.

[17] See IPCC (1994, p. 6).

next few decades. Indeed, some of this expansion will probably consist of the return into cultivation of land which has previously been set aside. Most of the required increase in global food production will come from more intensive use of land already in cultivation (i.e. higher yields). This leads us directly to issues of land quality.

Human-induced land degradation involves a host of complex processes ranging from soil erosion and pollution, to desertification. It is a serious problem with potentially significant implications for agriculture. The most systematic evaluation of the issue for the world as a whole is the Global Assessment of Soil Degradation (GLASOD) project which was sponsored by the United Nations.[18] It estimates that over the period 1950–90 about 17 per cent of the world's entire vegetated area lost some of its soil productivity. This figure consists of 6 per cent where degradation is judged to be light (productivity can be restored through farm conservation practices); 8 per cent where degradation is moderate (soil productivity is greatly reduced and needs major improvement to be restored); and 3 per cent where degradation is either severe or extreme (restoration would have to be large-scale or is practically impossible).

The GLASOD study distinguishes three main types of human-induced causes of soil degradation – all of roughly equal importance, and all closely related to food production. The first is overgrazing by animals, which is especially widespread and problematic in dry regions like the African Sahel and parts of Australia. The second is deforestation – usually for purposes of agriculture or (still more) large-scale commercial logging – which, for reasons elaborated below, can induce rapid loss of soil fertility. The third cause consists of various harmful agricultural practices, like faulty irrigation drainage or insufficient periods of fallow. A fourth, lesser, cause of degradation is the denuding of land for fuelwood – another process linked to population growth, and quite widespread in the Sahel.

Once the conditions for degradation have been established water erosion (primarily due to rain) is likely to be the single most important ensuing process – accounting for an estimated 56 per

[18] The following discussion of the GLASOD results draws on the accessible summary and map contained in World Resources Institute (1992, pp. 111–18). More detailed GLASOD maps, giving particular attention to Africa, are available in United Nations Environment Programme (1992).

cent of world land degradation. Next in importance comes wind erosion (28 per cent) which is particularly prominent in arid regions. Other significant agents of degradation include chemical processes (e.g. salinization due to poor irrigation drainage) and physical deterioration (e.g. soil compaction caused by heavy animals or machines).

Maps based on the GLASOD research are immensely detailed and intricate. But they highlight the following ten major world areas where degradation is already judged to be of serious concern:[19] (i) large parts of the United States Mid-West, (ii) much of Central America and the Caribbean, (iii) the Andean foothills and several parts of Brazil, (iv) most of South Africa, (v) virtually all of Madagascar, (vi) much of the Sahel – stretching from Senegal to Ethiopia, (vii) most of Turkey, and contiguous areas stretching south to the Persian Gulf, (viii) huge parts of the former Soviet Union (e.g. much of Ukraine and southern Russia), (ix) most of western India and Pakistan, and (x) much of eastern and northern China.

These GLASOD research results are certainly disturbing. But several points must be borne in mind when considering their implications for future food production.

First, it is an extremely difficult and controversial matter to gauge the extent of world land degradation. Even today the data-base for such an exercise is weak. And in 1950 – the study's reference baseline year – information on soil conditions was simply unavailable for most of the world. The GLASOD study therefore relies heavily upon the judgements of over 250 soil scientists. But the recent past provides evidence that expert opinion can sometimes overestimate the extent of degradation, and underestimate the recuperative powers of land.[20]

Second, it is important to appreciate that soils which are being eroded from one location are sometimes in effect being redistributed to other places. Therefore large soil productivity losses in one specific location – which may be comparatively easy to identify and measure, and which may indeed adversely affect crop yields there – may nevertheless be at least partly offset by marginal productivity

[19] The term 'serious concern' relates to widespread moderate soil degradation, or localized severe or extreme degradation, or some combination of both. See World Resources Institute (1992, p.117).

[20] Desertification in the Sahel has sometimes been exaggerated. Data provided by satellites in the 1980s do not support a general underlying retreat of vegetation. See Crosson and Anderson (1992, pp. 31–2) and World Bank (1992, p. 55).

gains over a far greater area.[21] Such considerations are certainly relevant within many of the GLASOD areas of serious concern.

Third, and most importantly, for several reasons it is difficult to see that the GLASOD estimates of land degradation will have much relevance for future aggregate crop yields and levels of world food output. The consequences for agriculture of a given degree of degradation depend critically upon the initial patterns of land use and soil productivity. A significant fraction of the land classified by GLASOD as 'degraded' lies in parts of the world that are of only marginal importance for aggregate levels of global food production.[22]

Moreover, soil degradation occurs in a *dynamic* context in which conservation measures may be instigated and fertilizer inputs raised.

As already intimated, studies in specific locations often show that soil degradation leads to falling yields. But it is very hard to demonstrate a convincing causal link of this kind for large territories. Crosson and Anderson note that those parts of the US Mid-West which GLASOD classifies as suffering from 'moderate' degradation – and therefore supposedly in need of major land improvement – have nevertheless experienced steady rises in cereal yields since 1950 because of the introduction of new crop varieties and higher farming inputs.[23]

Likewise, the previous list of ten major world areas which the GLASOD study suggests are of serious concern as regards soil degradation shows little correspondence to countries experiencing falling yields. Table 5.1 lists all sample countries where average cereal yields have recently declined. Perhaps only for Mexico, Cuba and Haiti is there any real correspondence to the soil assessment maps – and this is probably largely coincidental.[24] It is striking that

[21] Of course, such diffuse gains are much harder to measure. For a rare study of such soil redistribution in the US Mid-West see Larson *et al.* (1983).

[22] An obvious case is Iceland, which produces very little vegetative food – although it is classified as containing large areas of land of 'serious concern' for soil degradation. A similar point applies to much of the Sahel which is also classified as being of 'serious concern'. Even if there were to be no degradation at all in the Sahel the region's total volume of food production would be trifling in global terms (although obviously this is not to deny that degradation has adverse effects for the Sahel's population).

[23] See Crosson and Anderson (1992, p.34).

[24] It is true that much of Romania and the former Yugoslavia are classed as of 'serious concern' for soil degradation by the GLASOD study. But the inclusion of these countries in Table 5.1 is due to socio-political upheavals during 1987–92. For example, in former Yugoslavia cereal yields reached the record level of 4.63 tons per hectare in 1991, but plummeted to 3.15 tons with the start of hostilities in 1992.

Table 5.1 Sample countries with declining cereal yields between 1981–86 and 1987–92

Region	*Countries*
Sub-Saharan Africa	Mozambique (−39), Angola (−19), Malawi (−15), Somalia (−9), Zaïre (−8), Zimbabwe (−7), Niger (−4), Rwanda (−3), Ivory Coast (−2)
Middle East	Iraq (−6), Syria (−4)
South Asia	Afghanistan (−8)
Far East	Myanmar (−4), Malaysia (−1)
Latin America	Mexico (−12), Cuba (−12), Haiti (−8), Colombia (−2), Ecuador (−1)
Europe/FSU	Romania (−21), Yugoslavia (−7), Bulgaria (−4)

Note: Figures in parentheses are percentage declines. All other sample countries experienced increasing yields

most sample countries which are located in the GLASOD areas of serious concern (e.g. South Africa, Madagascar, Turkey) do *not* appear in Table 5.1. This lack of correspondence between judgements about soil degradation and falling yields is particularly pronounced for sub-Saharan Africa; note that only one Sahelian country – Niger – is listed. This serves only to emphasize that many factors affect yield trends. However, even a glance at Table 5.1 reveals the influence of instability and war (Mozambique, Iraq, Afghanistan . . .).

To conclude, as we do, that land degradation is unlikely to prevent a steady rise in global yields and food production in the period to 2020 is certainly not to deny its probable continuation and significance in some locations. We noted in chapter 4 that much of sub-Saharan Africa's increased food production will come from area expansion and shorter fallow. There, and to a lesser extent in parts of Latin America and South-East Asia, resource poor farmers will certainly continue to expand into tropical rainforests.[25] The luxuriant vegetation of these forests is the product of a rapid nutrient cycle, based largely on surface feeding, and average levels of soil fertility are rarely high. This cycle is effectively broken when the forest is cleared for cultivation. Organic nutrients, formerly preserved in the forest canopy, are broken down by high

[25] This discussion draws upon Duddin and Hendrie (1992) and Gleave (1994). Vivid descriptions of the causes and consequences of land clearance by poor cultivators are given in Harrison (1992).

temperatures, and rainfall leaches them away. Soils become friable. Their structure and nutrient status deteriorate, and a cycle of increasing erosion and falling yields sets in. After a few years the farmers are forced to clear new areas. Dryer savannah environments – notably the Sahel, but also parts of the Middle East and central Asia – are subject to broadly analogous processes.[26] The fact that, almost by definition, this kind of land degradation happens at the frontier of cultivation – where population densities are usually very low – makes it especially hard to control. The longer run agricultural, environmental and social consequences of these processes are often severe. But in the short term people's meagre crop yields, and therefore livelihoods, may be maintained.

Turning to consider the world's current stock of cropland, various forms of land degradation will certainly continue to occur. Especially noteworthy is the growing body of evidence that intensive irrigated rice cultivation over many years in parts of Asia has led to declines in underlying soil productivity – declines which have been masked by productivity gains arising from rapid changes in farm inputs and technology. This issue is certainly significant and probably relevant to other crops and locations. It seems likely that farmers are going to have to pay greater attention to care of their soil resource-base in the future. However, there is little doubt that such declines in soil productivity can be anticipated (this is one benefit from long-term agricultural experiments), treated and reversed (especially through information-intensive soil management procedures) and, most importantly for present purposes, more than offset by further changes in technologies and inputs.[27]

So, in many countries such processes are unlikely to have a large negative effect on yields compared to the considerable productivity gains which are potentially realizable simply from greater and more efficient fertilizer use. In fact, yield losses from land degradation tend to be greatest on tropical red-brown soils – which are often closely associated with rainforests and savannahs. But a disproportionate share of the global cereal harvest comes from black and brown earths in the temperate zone, and there is considerable

[26] On marginal dryland cultivation, and consequences like declining soil fertility, increasing drought and continuing poverty, see Glantz (1991).
[27] On these soil productivity issues see, for example, the contributions to Barnett *et al.* (1995). See also Cassman *et al.* (1993). In the mid-1990s a good example of a major offsetting technological change is the introduction of hybrid rice into countries like India and Viet Nam.

evidence that yields on these soils are fairly robust to light and even moderate degrees of degradation.[28]

Lastly, it must be emphasized that cropland can usually be rehabilitated if it is judged to be economically or politically worthwhile. Many kinds of action are possible. National governments may need to initiate large-scale programmes, such as watershed management schemes. At lower administrative levels, procedures like agroforestry, contour cultivation, crop rotation, mulching, manuring (and many more) can all be relevant. Soil conservation policies and programmes – until recently mainly the preserve of the United States – are now beginning to be adopted in developing countries.[29] In established agricultural areas of the world it is obviously easier to get new ideas and practices out to farmers in the fields. This said, at low selling prices farmers may see no point in land conservation, or they may simply prefer to raise their fertilizer use. The apparent presence of considerable supplies of unexploited fertile land, even in some of the world's most densely settled rural areas, suggests that often there is little real land shortage, let alone a need to restore degraded land.[30] In addition, currently there are significant reserves of cropland idling in North America and Europe.

Water for agriculture

There is widespread agreement that the future supply of water for agriculture represents a much more significant constraint to raising food production than do any likely foreseeable difficulties relating to soil or land.[31] It is frequently said that the expansion of global irrigation capacity is now slowing.[32] And many experts seem to

[28] See Crosson and Anderson (1992, pp. 15–43) and Duddin and Hendrie (1992, pp. 32–41).

[29] During the 1980s, with Swedish help, Kenya's national conservation campaign terraced about 40 per cent of the country's farms. See Hendry (1991, p. 28).

[30] On supplies of unexploited fertile land, see Smil (1994, pp. 275–7). Since 1950 China has lost about 20 per cent of its arable area, but about 5 per cent of the country's cropland is estimated to be idle.

[31] For example, Smil (1994, p. 277) writes that 'Agriculture's water needs will pose a greater supply challenge than land availability.' See also Alexandratos (1988, p. 132) and Crosson and Anderson (1992, p. 54).

[32] For example, according to Conway *et al.* (1994, p. 20) 'in recent years there has been a significant decrease in the rate of expansion of irrigation as real costs of irrigation projects have risen.' See Postel (1992, pp. 50–2; 1994, p. 44), and World Resources Institute (1992, p. 98).

agree that in the next few decades the world's irrigated area is unlikely to increase at the same pace as in the past.

Yet, especially in the Far East and South Asia, most of the required increase in future food production will have to come from the irrigated area – which already produces more than half of the food in these regions. Major agricultural areas like northern China and western India even now face acute seasonal water shortages. The situation is particularly urgent, even precarious, in the Middle East.[33] In all these regions, and indeed elsewhere, underground water reserves are often being overpumped in order to supply the agricultural sector, and water tables are falling. Desalinization of sea water for agriculture is much too expensive to be widely employed. Water scarcity is just as acute in eastern and southern Africa, where virtually all food production is rainfed.[34] Only in North America/Oceania and Europe/FSU – predominantly temperate regions, where the bulk of food is also cultivated on rainfed land – can we perhaps be reasonably confident that water problems will not constitute a hindrance to future growth in food output. And even these regions sometimes experience droughts, and contain areas – like the south-western United States – which seemingly face serious water shortages for irrigation.

This said, there is also widespread agreement that world agriculture is currently highly inefficient in its use of water resources. In all world regions there is scope for growing a lot more food from existing water supplies – given appropriate technical, management and institutional changes.

Estimates made in the early 1990s suggest that world agriculture accounts for roughly 69 per cent of total human freshwater use, with the industrial and domestic sectors accounting for 23 and 8 per cent respectively.[35] Large-scale, government irrigation systems – based mainly on age-old techniques of gravity flooding – are the prime users of water in the agricultural sector. These systems are usually poorly maintained and managed. They generally deliver water to the fields either free, or at very little cost to farmers. The water often arrives in the wrong quantities, or at the wrong times. Seepage and poor drainage frequently lead to problems like waterlogging or salinization of fields. And when, for example, it is

[33] See World Bank (1993c, p. v) which also talks of an 'impending crisis' in this region.

[34] See Falkenmark (1991).

[35] See World Resources Institute (1994, p. 346).

stated that 8–12 per cent of the world's irrigated area is 'seriously salinized' (figures which, inevitably, have to be regarded with caution) it is overwhelmingly to large-scale, government-run, irrigation systems that such estimates pertain.[36]

The main solution to the growing problem of water scarcity is generally agreed, easy to state, though considerably harder to implement: ways must be found of raising the *price* of water, especially for farmers operating within large-scale irrigation systems. The benefits of this approach are broadly threefold. First, it should improve the efficiency with which farmers use water, and reduce wastage and its associated problems (e.g. waterlogging, salinization). Second, it should increase the investment resources which are available, both for the development of new irrigation capacity (the costs of which are generally rising) and for improved maintenance of existing capacity (which may often be a more cost-effective approach than building new irrigation). Third, raising the costs of water to farmers, and therefore the efficiency with which they use it, is one way of releasing supplies for the growing industrial and domestic (urban) sectors.

Worldwide, it may be that under 40 per cent of the water which is currently diverted for irrigation purposes actually reaches any crops.[37] So even modest efficiency gains apropos the 69 per cent of freshwater withdrawals that are currently used by agriculture should often be sufficient to meet alternative demands. Already the pricing and trading of water between urban and rural sectors is becoming commonplace in southern and western states of the United States. Another approach is the pricing and re-use of nutrient-rich urban wastewater for irrigation – as, for example, around Mexico City. Although the process has a long way to go, the 1980s and early 1990s have seen the rapid emergence of innovative water-pricing and marketing schemes in areas of shortage around the world. Sometimes water markets have arisen quite spontaneously – as in parts of India and Pakistan, where farmers with deep tubewells now compete to service the water requirements of farmers without. But there is little doubt that govern-

[36] See Postel (1994, p. 44). She cites estimates that salinization affects 28 per cent of the irrigated area in the United States, 23 per cent in China, 21 per cent in Pakistan, 11 per cent in India and 10 per cent in Mexico. Although salinization tends to lower yields, it should be noted that some cereal crops like barley, wheat and sorghum are comparatively salt-tolerant.

[37] See Postel (1992, p. 100). See also Smil (1994, p. 278).

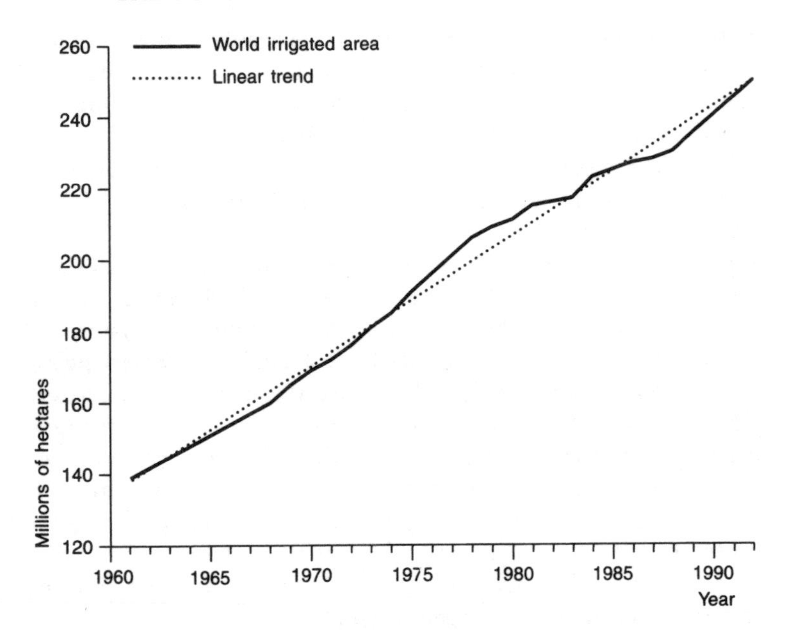

Figure 5.2 Growth in world irrigated area, 1961–92
Principal source: Postel (1994)

ments will have to become increasingly prominent actors in restructuring patterns of water-use incentives during the coming years.[38]

It is also important not to underplay the prospects for further expansion of the world's irrigation capacity. It is perfectly true that in percentage terms the average annual rate of global irrigation growth has been declining. But as Figure 5.2 shows, the general pattern of increase over time can be considered as being close to a linear trend. In the early 1990s the world's irrigated area was still expanding at about 3.6 million hectares per year – similar to the long-run average annual rise.[39]

[38] For example, in Chile the government has used the legal system to set up tradable water rights throughout the country; see Rosengrant (1993, p. 4). Where irreplaceable underground water reserves are being depleted governments may need to restrict withdrawals by taxation, or by raising electricity pumping costs. For more on these issues see Postel (1992).

[39] The point being made here is analogous to that made in respect of cereal yield growth in chapters 3 and 4.

As might be expected, most of this expansion is happening in the Far East and South Asia. In fact, for the world as a whole the ratio of irrigated area to population in 1991 was almost identical to the figure in 1961 – at about 45 hectares irrigated for every thousand people.[40] If the global irrigated area continues to expand at its long-run linear rate then by the year 2020 it will cover approximately 350 million hectares, and the ratio per thousand population will fall (very slightly) to around 44 hectares. Such a scenario cannot be ruled out. According to World Bank estimates the remaining potential irrigable area worldwide is about 137 million hectares – with roughly half of it lying in Asia.[41] If this figure turns out to have any validity then we might anticipate that there will be some diminishment in the annual linear expansion of irrigated area before the year 2020.[42] However, past experience in estimating apparently finite limits – such as the size of global mineral reserves – suggests that we should be cautious in placing confidence in any single figure limit.

In the late 1980s there were at least 1026 large dams (i.e. over 15 metres in height) being constructed in the world. Of this total, China and India alone were respectively building 183 and 160 dams which were over *30* metres in height.[43] While, of course, dams are also built in order to generate hydro-electricity, these figures are eloquent testimony to a massive and continuing commitment to raise irrigation capacity in Asia. Because of increasing attention to environmental and social costs, future dam and irrigation schemes may sometimes be on a somewhat reduced scale. But many huge projects are still being planned and executed.[44]

Furthermore, statistical data on irrigation are hardly precise. An appreciable component of future irrigation expansion is likely to occur through the building of much smaller dams, and through the introduction of low cost water pumps and drip irrigation. Such

[40] See Postel (1994, p. 45).
[41] See the figures cited in Crosson and Anderson (1992, p. 47).
[42] But even on the basis of this 137 million hectare limit such a diminishment is not inevitable. Thus if the global irrigated area increases at 3.6 million hectares each year during 1990–2020 then the resulting total increase is 108 million hectares.
[43] For these data see World Resources Institute (1992, pp. 330–4). In 1950 China contained only 2 dams over 15 metres. By 1986 there were 18,820 dams over 15 metres in China, i.e. over half the global total.
[44] A few examples are the giant north–south water transfer project in Spain, the scaled-down but still very large Narmada Valley project in central India and the gigantic Three Gorges undertaking on the Yangtze river in China.

changes are not always captured by available statistics.[45] Nor should we forget to mention likely improvements in water conservation for rainfed agriculture. Developments here – significant in the lives of many poor farmers, if perhaps not in terms of world food production – include such things as the introduction of new mulching practices, and the use of stone lines or vetiver grass barriers to slow water run-off and increase yields.

To sum up, there obviously will be significant challenges in providing water for food production in some world regions during the next few decades. Though different in nature, the water constraints facing farmers in the Middle East and much of southern and eastern Africa are especially severe. Indeed, if water for irrigation were more plentiful in the Middle East then this region would probably not already be so dependent upon external supplies of food. And if levels of rainfall were greater, and more assured, in these areas of sub-Saharan Africa then the prospects for future food output and security there would certainly be better.

In many places around the world the effects of increasing demand for water will ramify across different aspects of life and pose serious challenges for prevailing institutions. Issues of 'hydropolitics' can be expected to become increasingly prominent – from the village level right up to the international arena in regions (like the Middle East and South Asia) where major rivers or underground water reserves straddle national boundaries. In many places water will certainly become an increasingly expensive agricultural resource.

However, apropos future food production prospects, it is important to place these problems in the broader global context, and not exaggerate them. In places where problems of water for agriculture exist there is usually considerable potential for efficiency gains. Moreover, as we have argued, in the period to 2020 there will certainly be considerable further expansion of irrigation capacity, particularly in the Far East and South Asia. And, perhaps especially in these two regions, many countries have the administrative and managerial capabilities needed to successfully manage water problems where they arise. Finally, the long-run expansion of food production for export in both North America/Oceania and Eur-

[45] For example, Postel (1992, p. 121) reports that when FAO adjusted its data for sub-Saharan Africa to include small-scale irrigation, the region's estimated total irrigated area increased by 37 per cent.

ope/FSU is unlikely to be much constrained by shortages of water – although there will be periodic droughts.

FARM INPUTS, CROP DEVELOPMENTS AND RESEARCH

If achieving steady rises in food crop yields is the central challenge facing world agriculture today, then success will depend critically upon farmers having access to larger quantities of conventional agricultural inputs (especially fertilizers) and greater access to both existing and currently developing agricultural technologies. From time to time there may be significant advances – for example, from research in biotechnology. But as we saw regarding the 'green revolution' in chapter 3, even very significant breakthroughs tend to become submerged and smoothed out in the overall process of agricultural advance. In the real world progress tends to be *incremental.* Research aimed at improving local land and water conservation procedures, or increasing marginal agricultural efficiencies, or improving the quality of agricultural extension services so as to better communicate with farmers, is hardly glamorous. But we can confidently expect that future increases in food production will often depend more upon such rather prosaic developments than upon any 'magic bullets'. We address some of these issues here.

Fertilizers

World fertilizer production increased tenfold between 1950 and 1990.[46] This was probably the single most important factor which enabled the rise in global food output over the same period. In the early 1990s there has been a significant (14 per cent) decline in world fertilizer use. This happened mainly because the countries of the FSU (especially Russia) reduced their fertilizer subsidies, as indeed did several other nations under the conditions of economic structural adjustment programmes.

Yet the future expansion of world food production will require big increases in global fertilizer use. There is no alternative. Synthetic nitrogen is easily the most important example.[47] In 1990 world consumption of nitrogen was approximately 80 mil-

[46] For fertilizer use statistics see Brown (1994c, pp. 42–3).
[47] The following discussion draws on Gilland (1993) and Smil (1991, 1994).

lion tons, i.e. about 15 kg per person. A minimum first estimate of requirements in 2020 – holding this per capita figure constant, and accepting the UN medium variant population projection – would therefore be around 120 million tons. However, this estimate makes no allowance for a likely rise in indirect cereal demand caused by probable increases in the consumption of livestock products. It also makes no provision for the fact that there are diminishing yield returns to increased fertilizer applications.

Therefore probably a more realistic approach is to assess the amount of nitrogen required to produce our 2020 cereal production estimate of nearly 3 billion tons, and then make an additional allowance for the production of non-cereal foods. Such an approach suggests that agriculture will require about 155 million to 170 million tons of nitrogen annually by the year 2020, i.e. roughly a doubling of the world's present production capacity.[48] Of course, as with water, so with fertilizers, there is scope for efficiency gains. But the evidence seems to be that unless there is an unexpectedly rapid shift away from using urea, this scope is only moderate.

Fortunately, a doubling of world synthetic nitrogen production is feasible. Supplies of atmospheric nitrogen are virtually limitless; and there is no shortage of either methane or coal, which are both used in the production process. With the exception of sub-Saharan Africa, the capacity to manufacture nitrogen fertilizers is widespread in most world regions. Throughout the developing world much of this current capacity is in the form of under-sized and inefficient plants, which are badly in need of modernization. However, we can anticipate that the next few decades will see a continual upgrading and expansion of this nitrogen production capacity, especially in larger developing countries. China is already the world's largest single manufacturer. And on the basis of past trends it must be considered virtually certain that India will become the second producer – overtaking the United States – some time in the first decade of the twenty-first century.

Nor are there major constraints to increased production of phosphate and potash. There are ample mineral reserves for both these fertilizers – although it should be noted that commer-

[48] For the cereal output figure of 2955 million tons see Table 4.7. The range of 155–70 million tons is loosely based on Gilland's (1993) linear and parabolic equations relating world cereal production to total synthetic nitrogen consumption. This approach makes some allowance for decreasing yield returns.

cial supplies of phosphate rock are disproportionately located in the Middle East (about 50 per cent) and South Africa (about 25 per cent).[49]

In summary, it is worth quoting Smil's assessment that 'fertilizer production is a mature industry steadily improving its techniques and lowering its energy costs'.[50] It is hard to underestimate the relevance of this for the prospects of raising world food output, at reasonable cost and in line with the growth in demand.

Pesticides

The production of pesticides will be less important for future growth in global food production. Nevertheless, it is worth making a few comments about pesticides in order to illustrate the changing nature of much agricultural research and progress.[51]

The long-run trend in world consumption of pesticides has broadly paralleled that of fertilizers. In the early 1990s levels of pesticide use fell in many countries, as government subsidies were reduced. In addition, there are growing concerns about the harmful agricultural, environmental and health consequences of pesticides. And, unlike fertilizers, there are alternatives. These alternatives go under the title of 'integrated pest management' (IPM) schemes. They control pests by combining detailed research-based knowledge about crop–pest dynamics with a variety of non-chemical procedures – like crop rotations, ploughing-in stubble and the use of pest traps. IPM is a growing worldwide movement. In the late 1980s Indonesia provided a textbook case of how such procedures, spread through a major farmer education programme, could greatly reduce brown plant hopper damage to the country's rice crop – while simultaneously raising output and reducing costs.

Looking to the future, an increasing number of the larger developing countries will expand their own pesticide production capacity – which currently is overwhelmingly located in North America and Europe. Doubtless world pesticide use will continue to grow – but not as inevitably as will be the case with fertilizers. Instead, pesticide use can be expected to become much more

[49] See Mitchell and Ingco (1993, p. 70).
[50] See Smil (1994, p. 274).
[51] This discussion of pesticides draws on World Resources Institute (1994, pp. 111–18) and essentially subsumes herbicides and insecticides.

selective. And IPM provides a significant illustration of the kinds of incremental, grassroots, research-based procedures whereby food production will be intensified and costs cut in the decades ahead.

Foodcrops and biotechnology

From mung beans to maize, improved varieties of many crops are continually being developed and introduced. The promise of biotechnology gets most of the newspaper headlines. But, perhaps especially apropos the world's developing regions, conventional plant breeding programmes are likely to remain the main source of improved foodcrops, at least for the next two or three decades.[52] This is a huge subject. But some of the more promising developments are worth mentioning by way of illustration.

Particularly noteworthy are the new generation of rice varieties which are currently being developed in China and at IRRI in the Philippines. The new inbred rices have characteristics like very sturdy stems, vigorous roots, few unproductive tillers and three or four panicles (each with 200–50 grains); they promise yield improvements of 30 per cent compared to previous semi-dwarf rice varieties. Another major advance is the development of hybrid rices for the tropics. When the new inbreds are eventually subject to hybrid rice technology (so gaining hybrid vigour, a universal feature in biological systems) the combined effect may raise rice yield ceilings by up to 50 per cent.[53] These developments could make a very significant contribution towards raising the average cereal yield in the Far East to the level of 6 tons per hectare which is envisaged by the scenario in chapter 4. And, of course, rice yields in many other countries (e.g. Egypt, India) will benefit too.

Also noteworthy are the new varieties of triticale – a cross between wheat and rye – which are currently being developed at CIMMYT in Mexico.[54] Triticale is a fast-growing protein-rich cereal, with significantly higher yields than wheat, and with considerable tolerance to both salt and drought. It is being increasingly

[52] In this context it is worth quoting Vernon Ruttan's statement that 'Advances in conventional technology will remain the primary source of growth in crop and animal production over the next quarter century.' Cited in Brown (1994a, p. 187).
[53] See, for example, IRRI (1993) and Virmani *et al.* (1994).
[54] See World Resources Institute (1990, pp. 98–9).

grown in Australia, France, Poland, Mexico and parts of the former Soviet Union.

Lastly, we can mention new cultivars of pearl millet, which have greatly improved resistance to mildew disease and higher heat-tolerance than previous varieties. Developed at the International Crops Research Institute for the Semi-Arid Tropics (ICRISAT) in India, these new millets produced twice the yields of other varieties during the severe drought in sub-Saharan Africa in 1992.[55]

Like many other promising crops, these new cereal varieties are being developed at major agricultural research centres – notably those of the Consultative Group on International Agricultural Research (CGIAR) – in collaboration with expertise from a global network of national research institutions and universities. It is essentially this *public* network which has generated much of the basic research behind many of the food productivity gains of the developing world during recent decades.[56]

In contrast, the leading centres of research in biotechnology are often located in *private* companies, chiefly in North America and Europe. Their research is mostly directed at commercial foodcrops which are destined for developed country markets. This fact, plus the inevitable time-lag (e.g. 10–15 years) which may be involved between initial research investment and a crop's widespread cultivation, helps explain why it will probably be a couple of decades before bioengineered staple foodcrops are being extensively grown in many developing countries. In any case, it seems judicious to be cautious about the gains which are promised by any new technology.[57]

None the less, the eventual outcomes of biotechnology for food production in the developed and much of the developing world

[55] See CGIAR (1992, p. 9).

[56] In 1994 the CGIAR consisted of 17 international research centres, including IRRI, CIMMYT, ICRISAT, ICRAF and IFPRI. Other CGIAR centres focus on topics such as aquatic resources management, livestock, animal diseases, plant genetic resources, irrigation, and the strengthening of national agricultural research systems. Despite its linchpin status and criticism of its costs, the CGIAR accounts for only 3 per cent of global public agricultural research spending. In the late 1980s its budget was kept roughly constant in real terms. See Conway *et al.* (1994).

[57] This caution is properly reflected in the titles of publications such as the question mark in 'Agricultural Biotechnology – The next "Green Revolution"?' – see World Bank (1990); see also Messer and Heywood (1990). Another relevant consideration is that consumers in North America and Europe seem to have doubts about bioengineered foods.

may be considerable.[58] Techniques using monoclonal antibodies and genetic engineering may some day produce crops which make more efficient use of synthetic nitrogen, have improved resistance to diseases and pests, have better storage qualities and are more nutritious. Foodcrops which already feature prominently in such research include tomatoes, potatoes, maize and soybeans. Biotechnology will provide benefits for livestock and aquatic production too.

The more technologically advanced developing countries like China, India, Indonesia, Brazil and Mexico are developing their own biotechnology capacity. And, despite a sometimes difficult commercial operating environment in which even minor research findings are routinely patented, the CGIAR centres and related institutions are now embracing this technology too. Furthermore, a handful of private companies and philanthropic foundations are trying to ensure that biotechnology will benefit some major developing country crops like rice and millet.[59]

It is inevitable that new crops which have benefited from biotechnology will often face problems when they are introduced – just as was the case with earlier 'green revolution' varieties. But biotechnology is likely to represent an eventual net gain for food production in most world regions. As is so often the case, the real challenge posed by this growing field of research is how to protect the interests of resource-poor small farmers, perhaps especially those in sub-Saharan Africa. The problem is not just that research aimed at their specific needs is relatively rare. It is also that biotechnology will probably lead to significant shifts in international comparative advantage which may be detrimental to the interests of poor countries. For example, advanced fermentation techniques in more developed countries could curtail the demand for products like cocoa, palm oil, soybean and vanilla which have hitherto been produced mainly in developing regions.[60]

To sum up, the message of this section is reasonably positive. The grounding for future world yield growth appears to be fairly firm. The general implications for food production costs of these

[58] For relevant reviews see Greeley (1992), Messer and Heywood (1990), World Bank (1990) and World Resources Institute (1994, pp. 118–125).
[59] In this context American and European companies like Monsanto and Zeneca, and the Rockefeller Rice Biotechnology Network, should be mentioned.
[60] For a discussion of possible changes in comparative advantages see Greeley (1992).

also 'human-induced' factors – like fertilizers, pesticides and higher yielding crops – are probably downwards. Fertilizers provide the best illustration of these points. But the basic argument extends over much wider terrain.

PHYSICAL AND HUMAN CAPITAL

In considering future prospects it is easy to become preoccupied with the immediate ingredients of food production – like cropland, water supplies and fertilizers. But it is vital not to neglect the broader context. Of course, many future developments cannot be foreseen. But from the experience of recent decades it seems reasonable to suggest that the general stock of physical and human capital – both of which condition the enterprise of food production – will continue to improve.

It will surely be the case that a higher proportion of the world's farmers will have access to electricity and machinery (e.g. tractors, threshers, harvesters) – probably at lower cost – by the year 2020. Likewise, a large fraction of the world's current harvest is subsequently lost due to spoilage and pests.[61] So, particularly in the developing regions, it is plausible to expect that storage facilities will often be improved, leading to a reduction in the scale of post-harvest losses.

This is an appropriate place to remark on the potential of synthetic plastics. Any observer of world farming today must be impressed by the tremendous, if diffuse, impact of cheap new forms of plastic. Plastic sheets help retain soil moisture and keep down weeds. Plastics allow the use of cloches and greenhouses on a scale which otherwise would be unimaginable. Drip irrigation depends upon plastic tubing. And plastics enable improved storage of food. In the future such benefits can be expected to touch the lives of many more farmers.

Another area where improvements can be expected is communications. The scenario in chapter 4 involves a roughly threefold increase in the volume of cereals transferred from North America/Oceania and Europe/FSU to the remaining world regions. It is difficult to believe that world shipping will not be up to this task.

[61] Ehrlich *et al.* (1993, p. 16) cite an estimate that 40 per cent of the global harvest may be lost in this way. Smil (1994, p. 271) states that cutting current world post-harvest losses by one-fifth would raise global food energy availability by 6 per cent.

And presumably road and rail communications within many countries will improve too. The possible ramifications for future food prospects of better communications are many and subtle. For example, other things being equal, improved transportation probably reduces post-harvest losses. Or, to give another example, satellite remote-sensing promises to give European and North American farmers detailed mappings so that they can tailor fertilizer applications within individual fields, cutting waste where other factors limit growth. And in sub-Saharan Africa satellites are increasingly helping to monitor rainfall, and predict the movement of pests (e.g. locusts and army worms).[62]

Turning to the stock of human capital, its quality too will doubtless improve, at least in average terms. In the 1980s and early 1990s there were tremendous rises in child immunization coverage in many countries, with associated improvements in health. One consequence of this should be a fitter agricultural workforce in the coming decades. Though real, the beneficial effects of such an improvement for future levels of food production are probably tiny and immeasurable. But the same is probably true of many other individual influences which, when combined or operating in specific locations, may have a significant effect.

Arguably much more important could be the farm productivity gains arising from higher levels of *education*. In the mid-1990s most developing countries are educating an appreciably higher fraction of their children than was the case two or three decades before. Therefore the likelihood must be that the world's farmers will be somewhat better educated in the year 2020 than is generally the case today.

Better educated farmers are probably more confident, more innovative and more precise in the communication of their problems. It is notable that just as higher levels of education are now widely regarded as a particularly potent route whereby levels of birth and death rates may be reduced in developing countries, so there seems to be an increasing recognition among agricultural experts that the expansion of rural education is a key way through which the future productivity of farmers, and their fields, can be raised. The adoption and development of more efficient and productive methods of food cultivation should be greatly facili-

[62] See Ottichilo (1993).

tated as levels of education – among farmers, agricultural extension workers and researchers – rise in the years ahead.

Admittedly, it is important not to get too carried away by all of this. It is worth recalling, for example, that the food production performance of sub-Saharan Africa in recent decades has been dismal despite general improvements in average levels of health and education. Nevertheless, at least at the world level, it seems reasonable to assume that the quality of both the physical and human resources informing and conditioning future food production will advance.

INSTITUTIONS AND MARKETS

Most of the projected increase in world food production during the period to 2020 will probably happen largely irrespective of the details of any changes to prevailing institutional structures. But such changes may well affect output at the margin, and influence the *distributional* basis whereby gains in food production – and consumption – are achieved. For example, while it appears inevitable that the developing regions will have to import much larger quantities of cereals in the future, the extent to which this happens will partly depend upon the investments which are made in their own agricultural sectors – and, in turn, this may be influenced by institutional structures.

In the mid-1990s the future is predominantly discussed in terms of 'restructuring', 'privatization', 'comparative advantage' and, above all, 'properly functioning markets'. This is true at most levels, from the national to the international.

Almost certainly, the next few decades will see a gradual 'liberalization' of international agricultural trade. This trend away from protectionism is already exemplified, for example, by the completion of the Uruguay Round of the General Agreement on Tariffs and Trade (GATT) in the mid-1990s and the creation of the World Trade Organization (WTO). Some may see the trend as reflecting the shift to the right in political attitudes which has occurred since the early 1980s. But, more realistically, it may simply be a largely inevitable outcome of the tremendous expansion of global transport and communications.

A key feature of liberalization is the move away from subsidies, towards greater reliance upon markets and prices. At the international level the claimed benefits of this approach include the idea that levels of subsidized food production in North America and (still more) the European Union will be restricted, and that therefore less food will be produced and subsequently 'dumped' on world markets. In addition, these two major producing blocs, and some other rich countries (e.g. Japan), will lower their import restrictions bit by bit. In theory, international food prices will rise, and food producers in the developing world should benefit, expanding their own production.[63]

It is clear that broadly analogous changes are happening within many countries. For example, government-run agricultural organizations are being privatized and, as we have noted, subsidies for farm inputs like fertilizers are often being reduced. For numerous developing countries the implementation of structural adjustment programmes has involved the abandonment of economic policies which have hitherto had the prime *political* purpose of keeping urban food prices low. Such policies – like food price controls, and the deliberate overvaluation of exchange rates – have effectively worked against the domestic agricultural sectors of many countries. The expectation is that rising international, and national, food prices will benefit rural producers throughout the developing world.

Many developing countries – and also those in eastern Europe and the former Soviet Union – have effectively been forced by events in the 1980s and early 1990s to make these changes rapidly. But at the world level, although greater liberalization is virtually assured in future rounds of the GATT, recent history strongly suggests that it will be a comparatively protracted and piecemeal process. There are powerful political interests, and arguments, arraigned against any sudden liberalization. Most farmers in North America and Europe would certainly not welcome any abrupt change. Food is a special commodity, and many national governments have serious reservations about relying solely upon international markets for future supplies. Since most developing countries are net food importers, with rapidly growing urban populations, any sudden rise in food prices may harm the interests of more people than it helps (at least in the short run).

Unless the process of liberalization is moderated it will probably

[63] According to Pinstrup-Andersen (1994a, p.9), 'Available estimates suggest a 10–15 per cent increase in real agricultural prices on the world market if the original goals of the Uruguay Round [of GATT] are achieved.' For other estimates see Anderson and Tyers (1991) and Wyles (1993).

lead to widening differentials with respect to future food production, consumption and levels of food security. This is certainly a real danger. Preaching a message of comparative advantage is okay for those who actually *have* some advantage. But many countries and people may not. Indeed, contrary to the currently prevailing economic wisdom, the long-run outcome of trade liberalization may well reveal that most of the comparative advantage with respect to future food and agricultural production actually rests primarily with the farmers and associated industries of the developed world.[64] Such a conclusion can also be seen as one implication of the basic numbers we examined in chapter 4.

This brings us to the related issues of redistributive mechanisms and *public* action. Although one can only speculate, it seems plausible that in the coming decades these sorts of processes may generally be in retreat.

It is likely that the rather *ad hoc* coordination of surplus food stocks held in North America, and increasingly Europe, will continue to be the main source of emergency food assistance for poor countries during times of crisis. Notwithstanding the storage costs involved, there is no doubt that these reserves will continue to be maintained at fairly high absolute levels in the future. However, liberalization may well lead to some reduction in these stocks – at least as measured as a proportion of annual global production. Therefore, there may also be reductions in the average volumes of food aid which are routinely 'given' away.[65] This said, the needs of the very poorest countries are likely to be partly protected by the future stipulation of international commitments to provide food aid in quantitative volume terms – irrespective of prevailing world price levels.

Between 1980 and 1990 the amount of financial aid provided by developed countries to assist agriculture in the developing world declined from $US 12 billion to 10 billion (measured in constant 1985 dollars). Also, in the late 1980s and early 1990s spending on international agricultural research – of the kind represented by the CGIAR centres – has stagnated in real terms.[66] Although the theory behind trade liberalization is that everyone will be better off, and therefore developed countries will be able to devote

[64] This argument has perhaps been taken furthest by Carruthers (1993) in his appropriately titled paper 'Going, going, gone! Tropical agriculture as we knew it'.
[65] For these arguments see Matthews (1993). See also Figure 3.6.
[66] See Braun *et al.* (1993) and World Resources Institute (1994, pp. 110–11).

greater financial resources to international development assistance, one can surely be forgiven for expressing some scepticism that this is how things will actually turn out! Certainly, the foregoing facts do not augur well for future prospects for aid to developing country agriculture. And, as we have already noted, the private sector is unlikely to be very attracted by research investments which are chiefly aimed at helping poor farmers, and which pay back (if at all) only after many years.[67]

Perhaps partly as a consequence of these kinds of pressures, non-governmental organizations (NGOs) are likely to become increasingly prominent actors in the future agricultural institutional array – as innovators, facilitators and agitators across a whole range of activities. For example, they are already involved in applying biotechnology to traditional foodcrops which are otherwise being neglected.[68]

In conclusion, there must be a real risk that in the next few decades governments everywhere will find it expedient to lower their commitment to public action – under pressure to restructure, reduce taxes or cope with new problems. As a result, both within and between countries, differentials in food consumption and security could widen, despite the fact that liberalization may mean that average levels of food intake are somewhat higher than they would otherwise be. Widening differentials are already evident within major countries like China and India. And, at the international level, the relative food and agricultural prospects for sub-Saharan Africa – compared to the rest of the world – are unlikely to be enhanced by such trends.

FOOD PRICES

Given the volatility with which markets react to short-term difficulties we cannot completely discount the chance of another sudden rise in world food prices, as happened in 1972–74. Drought remains the most likely trigger, though a nuclear accident in a major grain-growing region is another possibility.

The real price of internationally traded cereals has been falling for well over a century. A sudden big reduction in the agricultural

[67] Some research aims conflict directly with private interests. For example, the development of apomixis in rice would free poor farmers from having to buy hybrid seeds from commercial companies. See Virmani *et al.* (1994).
[68] See Messer and Heywood (1990).

subsidies – such as those paid to EU farmers – would cause a rise in international prices. But such subsidies will be reduced only by degrees under the GATT and its subsequent rounds. To the extent that there will be upward pressure on cereal prices from liberalization in the mid and late 1990s, levels of production in countries like Australia, Argentina and the US should be stimulated to have a partly compensatory effect. In addition, given their economic and other difficulties, the food demand of Russia and neighbouring countries has recently fallen, other things equal, exerting downward pressure on international prices.

Calculations which suggest that future prices will rise considerably because of liberalization may pertain more to a static than a dynamic world. Over the long run the whole point of liberalization is to produce more food at lower prices. Moreover, most of the factors considered in this chapter probably have downward implications for future food production costs over the long term. Any transaction costs arising from climate change – or indeed the process of trade liberalization itself – are likely to be modest and diffused through time. In specific locations some production costs will rise, for example, due to growing problems of water availability. But, for the world as a whole, we have argued that future food demand will largely be met from currently available supplies of land and water. What global agriculture will require in order to raise its yields will essentially be greater, and more efficient, inputs of *energy*. And whether in the form of fertilizers, pesticides, electricity or farm machinery, the relative prices of such manufactured inputs seem set to decline.

From time to time there will be periods when international food prices rise appreciably – for example, perhaps in the late 1990s due to the GATT. And there is always the chance of ENSO-inspired worldwide droughts. But our general view is that in real terms prices will probably be no higher in 2020 than they were in 1990. And they may well be lower.

To this conclusion we add one caveat. It is that the volume of food demand in the Far East – and especially China – does not sky-rocket well beyond the upper levels which were entertained in chapter 4. The Far East will surely be the largest cereal importer in 2020. Indeed, it could require cereals equivalent to more than the total exports from North America/Oceania.[69] But if this region's

[69] See Table 4.7.

annual volume of cereal demand were to far exceed, say, 1.1 billion tons then, other things being equal, this would certainly have a major upward impact on global prices. But in our view this will probably not happen.

CONCLUDING REMARKS

This review of factors which will condition food production in the period to 2020 suggests that there are no insurmountable difficulties in raising world food output to meet the projected demand. Over this time-horizon (although perhaps not eventually) even global warming seems likely to be accommodated. At the aggregate world level there is more than sufficient cropland. And the same is true of water for agriculture – although difficulties are apparent in some regions. The scope to continue raising yields from greater use of farm inputs (especially fertilizers), better farm practices, improved crops, and rising levels of physical and human resources (especially improvements in education) appears to be adequate to the task. Overall, the potentials for increasing global food production in the next few decades probably outweigh the constraints – though farming seems set to become more management intensive.

There clearly will be increasing problems in certain respects and in certain locations. For example, land degradation will doubtless continue to occur in all world regions. And some regions face mounting troubles apropos water. The implications of such problems for food production will be greatest in the poorest regions which also face the largest projected demographic increases. Perversely, even the conjectured effects of global warming appear to favour agriculture in the world's more developed regions.

The conclusions from this chapter seem to be broadly consistent with a scenario in which the predominantly developing world regions require increased transfers of food from North America/ Oceania and Europe/FSU. These two developed regions, with mainly temperate climates, advanced agricultural systems and together containing about 40 per cent of the global stock of arable land, may actually transpire to have the better of any future 'comparative advantages' in food production.

Current expectations are that most of the predominantly developing regions will experience sufficient economic growth to be able to finance their growing volumes of food imports – although sub-Saharan Africa is an important exception. If we are right in suggesting that future international food prices will not be any higher over the long run then this will facilitate importation. But low food prices are problematic for many of the poorest developing country farmers – and, essentially, their basic difficulties seem set to remain.

6

EXPLORING THE FUTURE: REGIONAL VIGNETTES

INTRODUCTION

This chapter contains brief overviews for the seven world regions. We outline what has already been concluded about recent trends and future prospects. And, drawing on additional material, we elaborate on some of the main relationships between food production and population dynamics within each of the regions. The emphasis here is more on what can plausibly be expected to happen, rather than on what needs to be done; and certain regions clearly merit greater attention than others.

SUB-SAHARAN AFRICA

Sub-Saharan Africa faces formidable food problems. Its people are generally very badly fed. Levels of national food security tend to be exceptionally low. Per capita cereal and food production appear to have been falling since the 1970s, and these falls have only been partially offset by increased imports – including growing quantities of food aid. Therefore levels of per capita food consumption have probably declined slightly in recent decades. Certainly, this is what many people in the region *say* has happened.

The increase in the total amount of food being produced has mostly occurred through an expansion of the area being harvested. Average yields have probably not risen by much. Demographic growth has been the overwhelming cause of this area expansion – an expansion which has been accomplished by a combination of sharp reductions in the length of regenerative fallow and the clearance of new land. Often these processes have led to loss of soil fertility, erosion, etc.

In general, contemporary rapid population growth in sub-Saharan Africa has not resulted in the spontaneous development of higher levels of agricultural productivity. Nor, typically, has it resulted in a commensurate move towards more intensive farm technologies. Instead, in many locations farmers are exploiting the land for their own short-run survival, rather than husbanding it as a long-term productive resource.[1]

Furthermore, since the mid-1970s this is the only world region which has experienced major famines. Indeed, there have been quite a few – usually associated with warfare. In addition, there are distinct signs of climate change – manifested in more variable rainfall patterns and more frequent drought, especially in eastern and southern Africa.

The potential agricultural productivity of sub-Saharan Africa may well be high – certainly it is sufficient to meet the food requirements of the region's current population.[2] But this theoretical capacity has to be seen in the context of fundamental sub-structural difficulties like very rapid demographic growth, pervasive poverty, ethnically heterogenous nation states, poor systems of administration, political instabilities and, therefore, an extremely weak international comparative position.

A small part of the decline in per capita food production since the 1970s may have been due to increased cultivation of non-food cash crops for export – such as coffee, cotton, pyrethrum, tobacco and tea.[3] But the region has experienced increasing competition in these crops, and in the production of fruit and vegetables too, especially from large-scale Latin American producers who are better resourced and organized. Moreover, sub-Saharan Africa has faced generally worsening terms of trade, which has made the import of chemical fertilizers and farm machinery very difficult.

It is important to bear in mind that sub-Saharan Africa's dominant modes of agriculture are the most distinctive of any world region. This has many sides. For example, probably a majority of the region's subsistence farmers are women.[4] In many areas the

[1] Of course, there are exceptions to these generalizations. For a comparative study of these issues in six countries see Lele and Stone (1989).

[2] On the region's agricultural potential see Higgins *et al.* (1981) and Plucknett (1993, pp. 9–10).

[3] The sentence is phrased cautiously because there are powerful arguments that such export crops have not held back food production. See Jaeger (1992).

[4] Women grow perhaps 80 per cent of the food consumed by rural households. Men are more prominent in cash crop cultivation. See Alamgir and Arora (1991, pp. 87–92).

hoe is still the most important agricultural implement. In much of the region cattle are not closely integrated into crop production – for example, in terms of providing draught-power and manure. Large-scale irrigation is very rare; indeed, the potential to establish such irrigation in the future is apparently fairly limited.[5] Even the region's predominant foodcrops are relatively distinct – with roots and tubers being the staples in much of central and western Africa.

An important repercussion of this distinctiveness is that 'green revolution' advances have often bypassed the region. Recall that these advances hinge critically around the triple package of irrigation, chemical fertilizers and HYVs. Yet sub-Saharan farmers rarely have access to, or hope of, the first two items. Moreover, most HYVs are of rice and wheat – cereals which are rarely grown in the region. The continuing widespread cultivation of maize in eastern and southern Africa is partly because it is the main cereal crop for which modern varieties exist which, along with traditional sorghums and millets, is reasonably suited to the local agricultural conditions.

The relative neglect of research on sub-Saharan agriculture is beginning to change. For example, the CGIAR has started to shift its research emphasis towards this region – giving somewhat greater priority towards root and tuber crops, and the needs of small farmers who cultivate rainfed marginal lands. Nevertheless, total agricultural research expenditures for sub-Saharan Africa have probably increased more slowly than those for any of the predominantly developing regions used in this study.[6]

Then there are the problems stemming from government policies. There is widespread agreement that the region's farmers have often been badly treated, and that they have received insufficient incentives to produce food for their own domestic urban markets. Faced with falling international food prices, poor internal arrangements for transporting and storing food, and extremely fast population growth in major towns (which are often ports), many governments in sub-Saharan Africa have increasingly relied upon cereal imports and food aid. The argument is that governments have neglected to invest in their own agricultural sectors. The most spectacular cases have been those nations which, at different times,

[5] The upper limit for the region's irrigated area appears to be about 20 million hectares. See Crosson and Anderson (1992, p. 47) and Grigg (1993, p. 154).
[6] This statement certainly holds for the period since the early 1960s. See World Resources Institute (1994, p. 111).

have controlled valuable export commodities. For example, Zambia deliberately imported food during the 1960s when the price of copper was high. Nigeria used its oil revenues for similar purposes in the 1970s and 1980s, and by itself accounted for much of sub-Saharan Africa's general cereal import growth at that time. Today Botswana uses its diamond revenues to pay its farmers to cultivate sorghum (the traditional cereal grain) while the country's townsfolk prefer to eat maize, which has to be imported.

One promise of the market-oriented structural adjustment policies which – usually at the behest of the World Bank and IMF – have recently been introduced by many sub-Saharan governments is that food prices will rise and farmers will have greater incentives to grow food for domestic markets. Already there are claims that these economic policies have led to improved food production performance.[7]

This said, there are good reasons to be cautious when considering the nature and scale of the benefits which are claimed for these new policies. Hopefully they will produce long-term economic growth which will help sub-Saharan agriculture. But in the short term structural adjustment is more likely to result in further declines in living standards and food intake.[8] One can also ask how great has been the positive impact of such policies compared, for example, just to the negative impact of an exogenous event such as the severe drought of 1992. More important still, structural adjustment policies do not really purport to address most of the previously listed sub-structural difficulties which together have certainly strongly conditioned the region's miserable food production performance. Also, it is hard to deny that this poor production performance has helped cause the general rise in food imports and aid.

Only by somehow increasing attention to, and investment in, domestic agriculture can sub-Saharan countries hope to improve their levels of per capita food production and consumption. And,

[7] For example, Jaeger (1992, p. xi) writes, 'Since the mid-1980s . . . there have been improvements in both food and export agriculture, a reversal which has coincided with the implementation of policy reform programs by a substantial number of African countries.' For similar claims see World Bank and United Nations Development Programme (1989).

[8] Against the long-run upward trend, there appears to have been some moderation in the volume of cereal imports received by the region since the late 1980s. However, rather than indicating improved production performance, this may be a sign of increasing financial difficulties.

either directly or in a regulatory role, governments will *have* to be involved in key tasks like improving rural transport systems, extending rural credit facilities, targeting agricultural research, improving the quality of agricultural extension services and raising fertilizer use from the extremely low levels which currently prevail. Clearly, these are hugely difficult tasks. And outsiders have not always been wise in the advice they have given in the past. But one thing is sure: relying upon 'market forces' alone is a recipe for disaster.[9] Governments in sub-Saharan Africa are very aware that significant fractions of their urban populations simply cannot afford higher food prices, whatever the outside advice may be. And we have seen that by 2020 about half the region's total population will live in towns. So the hugely difficult dilemma posed by the rural and urban sectors will grow. For sure, government policies can improve matters. But in some locations such improvements may turn out to be essentially epiphenomenal compared to the adverse effects of the underlying determinants of the situation.

Despite rapid urbanization, rural to rural migration is probably still the most prevalent type of human movement in sub-Saharan Africa.[10] Looking to the future it is evident that there will continue to be rural out-migration from those areas which are experiencing demographic pressures. These locations include not only densely settled areas like the Ethiopian and East African highlands, but also sparsely populated arid and semi-arid tracts such as are found in the Sahel. Much of this migration is, and will continue to be, towards the major towns. But there is also very considerable movement in search of cultivable land – sometimes over long distances. The most likely destinations for this latter type of intra-rural migration include forested areas of Cameroon, Congo, Gabon and Zaïre. Indeed, flows into these countries are already established.[11] It is true that malaria and trypansomal infections (the latter causes sleeping sickness in people, and afflicts livestock too) can slow the volume of migration into rural areas where they are prevalent. But it is also the case that increasing human population density may itself reduce the prevalence of these diseases – for example, by disrupting vector insect populations (e.g. mosquitoes,

[9] For a fine statement of this view see Lele and Stone (1989).
[10] See Oucho and Gould (1993).
[11] See Russell (1993).

tsetse flies) as new patterns of land use and vegetation are established. Moreover, methods of coping with insects (e.g. traps) are continually being improved.

Some comments are also required regarding HIV/AIDS. At present this disease is largely restricted to a large contiguous belt on the continent's eastern side, which stretches from Rwanda and Uganda southward to Zimbabwe (and, increasingly, into parts of South Africa too).[12] Food production in this belt has probably been adversely affected already – though not to an extent that is easily detectable in national statistics. Nevertheless, it is clear that there are many ways in which the disease can disrupt food production and security. An adult death reduces the size of the family labour force. If people are sick then the productivity of their farm work is reduced. Even those people without the disease may spend less time in farm work – as they care for those who are ill, or attend (often lengthy) funeral events.

However, for several reasons the influence of the HIV/AIDS epidemic on per capita food output in sub-Saharan Africa may not be quite as damaging as has sometimes been supposed. First, it appears that the effects of the disease on the overall population age-dependency ratio may be relatively modest. That is, the ratio of children to adults in the total population may not change greatly as a result of the disease.[13] Second, there is growing evidence that in sub-Saharan Africa's particularly stressful economic and health conditions, the interval between the initial onset of AIDS symptoms and death is appreciably shorter than is the case elsewhere. In other words, the average period of outright sickness is relatively short.[14] Third, and potentially most important, there are growing reasons to suspect that HIV/AIDS may spread rather slowly from its main belt in eastern and southern Africa. The population of this belt shares particular characteristics – lack of male circumcision and an associated very high level of the genital ulcer disease, chancroid – which are not found in most of the rest of sub-Saharan Africa. Accordingly, there is tentative conjecture that a

[12] See Caldwell and Caldwell (1993b). Abidjan in Ivory Coast and its surrounding areas is the only other – and demographically much smaller – location in sub-Saharan Africa which is currently beset by a major HIV/AIDS epidemic.

[13] See United Nations (1993, pp. 60–1). Clearly, other things being equal, if this ratio were to rise sharply then we would expect the food situation to be made even worse.

[14] See, for example, Anzala *et al.* (1993) and Mulder *et al.* (1994).

truly major epidemic will not become established beyond the eastern belt.[15]

One positive development in the next few decades will probably be an improved situation in the regional economic giant that is South Africa. From being a significant cause of the region's food problems – notably through the export of warfare and instability to neighbouring countries like Mozambique – South Africa could now become a major influence towards improvement, especially by contributing to greater, and more balanced, regional trade.

South Africa's new development strategy places great emphasis upon agricultural development and land reform. The potential benefits from this include increased per capita food output, rural employment creation, reduced risks of ethnic conflict within the country and moderated flows of migrants towards the towns (there is a backlog from the days of apartheid, when such movements were restricted).[16]

It remains to be seen how far the new South African government will actually implement the strategy, and avoid the easy option of simply buying in food from overseas. Still, at least the country's role as a deliberate agent of havoc is gone. And, in time, food production and consumption in neighbouring countries – such as Zambia, and especially Zimbabwe (where agricultural price and marketing reforms do seem to be having some success) – are likely to benefit from a growing South African market and, indeed, the export of South Africa's considerable technical agricultural expertise.[17]

However, for most countries in other parts of sub-Saharan Africa there can be little confidence that per capita food production will increase in a sustained way during the period to 2020. And even in those few cases about which hope is sometimes tentatively expressed – like Ghana, Kenya, Senegal and Uganda – there will be not only a need for the conventional list of previously cited agricultural changes (e.g. greater fertilizer use, and better trans-

[15] For these arguments see Caldwell and Caldwell (1993b). This said, as we implied in chapter 4, the UN population projections may understate the future scale of mortality from HIV/AIDS.

[16] See African National Congress (1994) and Lipton (1993).

[17] However, the short-term effects for neighbouring countries of the changed conditions in South Africa may be less favourable if, for example, the numbers of migrant workers from these countries are reduced.

port, credit, research, etc.), but also a near-absolute requirement for socio-political stability.[18]

So even with economic policy reforms, greater external agricultural assistance and an improved situation in southern Africa, there must be a strong likelihood that sub-Saharan Africa will be unable to raise its food production to match its population growth. It is worth considering, for example, that Nigeria's 1990 population of roughly 96 million is projected to reach about 215 million by the year 2020, when it will still be growing at 2.2 per cent per year. Sub-Saharan Africa seems set to become increasingly dependent upon food imports, including food aid. Probably the 'rich world's self-serving altruism' will provide some of the needed food transfers,[19] plus technical agricultural assistance. But, even so, average levels of per capita food consumption in the region could remain constant or decline still further.

The figures in chapter 4's scenario do not seem implausible – with perhaps a quarter or (less likely) one-third of sub-Saharan Africa's total cereal consumption being met by imports (and aid) by 2020. It is little consolation that this range falls short of the enormous numbers which are sometimes touted.[20]

This unhappy scenario may not necessarily result in famines – that will depend crucially upon related events in the political arena (although it is hard to be confident in this respect). And there are those who still maintain that population growth will be beneficial.[21] However, our view is that slower population growth would materially enhance the region's capacity to deal with its food problems, especially over the longer run. And many agricultural administrators and researchers in sub-Saharan Africa fervently share this view.

THE MIDDLE EAST

The Middle East's population is relatively well fed – at least in terms of the average number of calories available per head. This

[18] In this study we have to draw the line somewhere. But for the argument that population growth contributes to such instability – e.g. in the dispute between Senegal and Mauritania in the Senegal River valley – see Homer-Dixon *et al.* (1993).

[19] The expression is from Smil (1994, p.281).

[20] See Pinstrup-Andersen (1994a, p. 21; 1994b).

[21] Yoweri Museveni (1992, p. 235) – who takes a keen interest in food issues – nevertheless still argues that 'A country like Uganda is presently under-populated. There are 17 million of us but this population is not big enough to cope with the resources we have.'

said, there is tremendous diversity. For example, estimated per capita calorie availability in Sudan is below the average for sub-Saharan Africa. But the corresponding estimates for Egypt, Morocco, Syria, Tunisia and Turkey are all fairly high.

Despite its rapid demographic growth, both cereal and food production in the Middle East have broadly kept pace. But the considerable extra demand for food which accompanied the oil price rises of the 1970s could be met only by increased imports. And, of course, nowadays most of the region's people are accustomed to these higher levels of per capita food consumption.

We have seen that around 1990 cereal transfers into the region accounted for roughly 30 per cent of consumption – a figure which rises to about 40 per cent if Turkey is excluded. Egypt is the world's single largest recipient of cereal aid – overwhelmingly sourced by the United States. But it is worth noting that even in Egypt such aid accounted for only about 6 per cent of national cereal consumption around 1990. And Egypt has recently made considerable progress in its food exports (notably of oranges, potatoes, rice and tomatoes) to Europe, which help to finance the much larger proportion of its cereal consumption which it buys in (around 35 per cent in 1990). As a proportion of their total consumption both Sudan and Tunisia received greater amounts of cereal aid than Egypt – and, again, the US was the prime source.

Turning to national food security, by the FSI measure used here Turkey's level of food security is high; indeed, Turkey has significant potential to export food, especially to western Europe, but elsewhere too. However, partly because they are judged to have low levels of socio-political stability, we assess Algeria, Sudan, Iraq and Yemen to be of low, or very low, food security status. Both Iraq and Sudan have tasted near-famine in the early 1990s.

Water constraints are the chief reason why the Middle East will be incapable of increasing its food production in line with growing demand during the coming decades. Indeed, the region's dependence upon imported food has been depicted as an indirect way of bringing in water (in the form of the imported food's water content).[22] Despite its huge investment in wheat cultivation – which involves massive pumping of underground fossil water

[22] Thus the ability to import cereals – 'virtual water' – eases the region's tight irrigation constraints. See Allan (1995, p. 42).

supplies – even Saudi Arabia is still a net importer of cereals (especially barley) and of food in general.

We have seen that around 1990 about 17 per cent of the region's cropland was irrigated. Most of this irrigation is in Egypt, Iraq, Sudan and Turkey. With the exception of the last country (where the Southeast Anatolia Project will greatly increase the irrigated area) the scope to further expand irrigation capacity is very confined. Moreover, salinization and waterlogging affect much of the Middle East's irrigated area to some degree.

So apart from Egypt, Iraq and Saudi Arabia, most major Middle Eastern countries depend upon rainfed agriculture for most of their food production. This helps to explain the great volatility in the region's harvest from year to year.

There are many imaginative ideas – involving both physical and social engineering – on how to improve water conservation and allocation in the Middle East.[23] But actual progress is often difficult and slow. Undoubtedly, there is a risk of conflict over the waters of the Tigris–Euphrates; and there will be difficulties too regarding the Jordan River and the Nile.[24]

Despite these problems it is important to recognize the increases in cereal yields and food output which have occurred in recent years. Fertilizer use is still low in most countries – even on a lot of the irrigated area – and the coming decades will probably see much greater use of chemical fertilizers. Moreover, some of the new generation of foodcrops being developed in Asia will benefit agriculture in parts of the Middle East (especially on the irrigated area). In addition, Egypt in particular has its own agricultural research capacity which will probably make a significant contribution to food output during the next few decades.[25]

Of course, the Middle East's diversity extends to national views regarding demographic growth. Countries like Egypt, Tunisia and Turkey have family planning programmes and falling birth rates. But Iraq and Saudi Arabia do not share such goals. Corresponding to these differences there will be large differentials in future population growth which, in turn, will have a huge impact on the differential growth of food demand. For example, Egypt's population is projected to grow by about 63 per cent between

[23] See Allan (1995) and World Bank (1993c).
[24] See Postel (1992, pp. 73–86).
[25] See World Bank (1993d) regarding the likely expansion of Egypt's agricultural production.

1990 and 2020. But the populations of Iraq and Saudi Arabia are forecast to rise by 115 and 136 per cent respectively.[26]

There seems little reason to question our previous conclusion that the Middle East's considerable future increase in food demand will be met only to a limited degree from increased production within the region. Much of any increase in the total volume of food grown in the Middle East will happen in Turkey, Egypt and perhaps Morocco. Some of this increase will be exported to Europe (creating greater competition there, especially for Mediterranean producers). By 2020 the Middle East will probably be meeting roughly half of its total food consumption through traded purchases from outside. Countries like Saudi Arabia and Iraq should find the financing of these imports relatively undemanding, even should the international price of oil remain comparatively low. Other countries like Egypt, Morocco, Syria and Tunisia will find the task harder – but surely feasible.[27] Food aid shipments into the region for strictly humanitarian purposes in 2020 could still be largely restricted to Sudan.

SOUTH ASIA

Levels of per capita food consumption are very low in most of South Asia.[28] Indeed, by some external nutritional criteria (which may not be entirely appropriate) the region could contain half the world's undernourished people. Around 1990 South Asia imported only about 4 per cent of its cereal consumption. Most of these imports went to Iran – a country with levels of food consumption, income and indeed general food prospects which are much more akin to those of the Middle East. The remaining imports mostly went to Bangladesh – much of it as food aid.

So South Asia is largely self-sufficient in food. With the exceptions of Afghanistan and Iran, most countries enjoy moderate degrees of socio-political stability (although there clearly are some exceptional locations – like Kashmir). These facts help

[26] As elsewhere the figures cited are the medium variant projections from United Nations (1994). See the Appendix.
[27] We cannot address the economic prospects of individual countries. But they include non-oil mineral exports (including phosphates for fertilizers), tourism and remittances from labour migration within the Middle East.
[28] More detailed treatments of the issues considered here are contained in Dyson (1993) and Greenspan (1994).

explain why we judge the region to have a higher level of food security than its widespread poverty and meagre diets alone might imply.

In recent decades the regional level of per capita food production has risen significantly. This largely reflects progress in India. But there have also been rises in per capita food output in Pakistan (especially in the 1960s), Iran and Sri Lanka too.[29] However, per capita food production in Bangladesh has generally fallen since the 1950s. And Afghanistan's recent experience has literally been disastrous.

The risk of serious and extensive famine in most of South Asia today is significantly lower than was the case, say, in the 1970s. For example, India has used its improved cereal production performance to modestly increase levels of per capita consumption, eliminate its dependence upon imports[30] and accumulate sizeable reserve food stocks. These stocks helped the country to avoid widespread famine and deaths in 1987, despite a very severe drought.

India's great agro-climatic diversity – which hindered the occurrence of a sudden transforming nationwide 'green revolution' in the late 1960s – nevertheless provides a measure of stability, and a multiplicity of circumstances in which some agricultural progress can be made. Parts of India (especially Punjab) experienced the same dramatic rises in wheat yields shortly before 1970 as were occurring in neighbouring Pakistan (see Figure 3.4). And in both countries the immediate demographic consequences of this elevated productivity were similar – agricultural labour migrated into both Indian and Pakistani Punjab. But subsequently in India the momentum shifted to HYVs of rice, which tend to be cultivated in other parts of the country. More recently still there have been significant increases in the planting of HYVs of sorghum and millet.[31] In general, since

[29] Sri Lanka is fairly unique in South Asia in that despite population growth it has been able to increase its supply of cropland per person; the explanation is that malaria eradication programmes in the 1940s and 1950s used chemical sprays to kill vector mosquitoes – and hence opened up large tracts of land for agricultural resettlement and irrigation. The most important irrigation project was the Mahaweli Scheme, which the government viewed as a way to relieve rural population pressures, create employment and increase food production.

[30] In 1951 India's net foodgrain imports were equal to 12 per cent of net production. See Tyagi (1990, p. 50).

[31] See Agrawal *et al.* (1993, p. 122). The absolute yield gains which can be expected from these rainfed crops are obviously rather limited.

the early 1980s India has experienced a modest acceleration of per capita cereal and food production.

This is not to deny that demographic growth has presented a major challenge to agricultural production in South Asia. Since the early 1970s the average size of farm holdings in India has fallen from 2.3 to about 1.4 hectares. A typical observation is that 'Perhaps the most dominant and somewhat intractable problem of agriculture is the stupendous marginalization of holdings under the pressure of population.'[32] This process has gone furthest in Bangladesh, where by 1990 the average area of cropland per person was only about 0.08 hectares.

Turning to the future, fortunately India's growth in food demand will be constrained by the fact that the country's birth rate has been falling since the 1960s. The population is forecast to increase by about 56 per cent between 1990 and 2020. Most of the country's future growth in food production will continue to come from the irrigated area – where there are distinct opportunities for better utilization, more multiple cropping and significantly higher yields.[33] But it is from much greater fertilizer use that future gains in food production will largely come. While India has good supplies of coal and natural gas, its fertilizer industry is highly inefficient. There is also increasing recognition that some areas of cropland require greater inputs of phosphate and potash, as well as nitrogen.[34]

Increases in food production in India's rainfed sector – always subject to the vagaries of the monsoon rains – will inevitably be modest, at best. Especially from dryland regions the 'long anticipated great "shaking loose" of the rural peasantry from the villages' towards the towns is likely to be pronounced.[35] In eastern India near-feudal political institutions and high population densities are likely to impede potentially advantageous developments – like the spread of minor forms of irrigation. And, throughout the country, economic liberalization may tempt farmers away from growing staple crops which form the basic diets of many poor people.

[32] See Agrawal *et al.* (1993, p. 117).

[33] The spread of tubewells in areas where there are already large-scale irrigation canals has occurred partly as a response to ineffective irrigation management. Studies indicate that farmers with tubewells can double the cereal yields of their neighbours who rely solely upon canal water. See Postel (1989, p. 14 and p. 39) and Shirahatti (1989, pp. 92–3).

[34] See Sarma and Gandhi (1990, p. 39) and Shirahatti (1989, p. 94).

[35] This quotation by Ashish Bose is cited in Skeldon (1984, p. 28).

The economic changes already underway in India may well lead to greater differentials with respect to food consumption.

However, despite such problems, there can be little doubt that India has the capacity to raise yields sufficiently to meet the projected demand arising from its population growth plus some modest allowance for increased per capita consumption. Holding the area harvested constant, a cereal yield of around 3.7 tons per hectare would be required – a figure which is below the average yield achieved by Indonesia around 1990.[36]

Bangladesh's prospects are sterner. But they may not be quite as dire as they are sometimes portrayed. Estimated levels of per capita food availability in Bangladesh are exceptionally low. Around 1990 the country was receiving about 10 per cent of its cereals as imports and aid. However, since the early 1980s cereal yields have been rising at a fair pace; and the country's birth rate is also falling more rapidly than was generally expected. Clearly, if levels of per capita food intake rise in the coming decades (which is to be hoped) then this will probably raise Bangladesh's import requirements. This said, the country's economic performance has also been marginally better than was generally anticipated and, if continued, may help to finance food imports. Again, much will depend upon raising fertilizer use. But given the impoverished basis from which it starts, Bangladesh's long-run food prospects may turn out to be somewhat less grim than is often expected – in no small measure due to its progress in slowing population growth.[37]

Finally among the region's demographic giants there is Pakistan. Since the early 1950s Pakistan has been largely self-sufficient in food, and recently there has been considerable diversification in the country's food production. In each decade since the 1950s estimated food output has grown faster than population. This said,

[36] See Dyson (1993). Indonesia's yield for 1990–92 was 3.857 tons. See World Resources Institute (1994, p. 292).

[37] The medium variant population projection in the 1994 UN revision puts Bangladesh's 2020 population at 185 million, compared to 209 million in the 1992 revision. See United Nations (1993, 1994). Holding per capita cereal consumption constant at 1990 levels, the 1994 projection implies that the country's cereal requirement would rise to 48 million tons in 2020. Assuming, for simplicity, no change in area harvested, were the average cereal yield to rise from about 2.5 tons around 1990 to 3.7 tons in 2020 (probably a feasible proposition given increased fertilizer applications and use of hybrid rices) then total output would be about 41 million tons, i.e. 85 per cent of requirements. However, if the 1992 projection were to apply then this proportion would fall significantly (on these simple calculations) to 76 per cent. See Dyson (1993).

Pakistan's demographic growth has been, and will continue to be, formidable. The UN projects a population increase of 115 per cent during 1990–2020. Whatever may be the longer term implications of this growth, there is every reason to believe that Pakistan has the capacity to meet the resulting rise in food demand to the year 2020, even with some increase in average levels of consumption. As elsewhere in the region, there is very significant scope for greater fertilizer use. And the country's irrigation systems – which cover about 80 per cent of its cropland – suffer from all the problems which plague government-run schemes. Better management and utilization of this irrigation could certainly raise yields greatly. The extent to which Pakistan retains its state of approximate food self-sufficiency during the next few decades will probably depend mostly on policy decisions, themselves conditioned by external factors like trends in the international price of food.

To conclude, the food prospects facing South Asia in the next few decades are not without some limited promise. Population growth rates are generally falling. Fertilizer use can profitably be boosted. Irrigation capacity can be better used. Of course, much will hinge on wider developments. Certainly, compared with the 1970s India's economic circumstances show signs of improvement. There seems to be greater diversity in everyday life, and even very poor people may now occasionally spend marginal earnings on non-food items – like a bus ride, or perhaps some soap – which previously they would have spent on food. These personal observations caution against becoming overly focused on edible sustenance.

In 2020 a great many of the region's people will still be very poor, and very poorly fed. But if, as the scenario in chapter 4 suggests, the volume of net cereal transfers into the region has risen from 12 million to 50 million tons we can be fairly sure that a disproportionate share of this rise will go to Iran. Bangladesh, Pakistan and India may not experience a rocketing increase in imports – at least as measured as a proportion of their consumption.

THE FAR EAST

Containing one-third of humanity, the Far East – and, within it, China – will be particularly influential in determining world food prospects. We have seen that this region has experienced strong recent increases in food production and consumption; and average levels of per capita calorie availability are quite high. Largely because of Japan – and increasingly China – the region is already significantly more dependent upon cereal imports than, for example, is South Asia. And virtually all of these imports are traded, i.e. they are not received as food aid.

Of course, there is tremendous variation within the region. Per capita food availability is very low in Cambodia, higher in Indonesia and higher still in Japan. In several countries agricultural production has been badly disrupted by instability and war; Cambodia (since the early 1970s) and Myanmar/Burma (since the early 1980s) are prime examples. Thailand is still a significant net exporter of cereals – overwhelmingly rice – although this situation may soon cease. Myanmar and Viet Nam are roughly self-sufficient, and sometimes export rice. But by 1990 most other countries in the region were net cereal importers. Japan, South Korea and Malaysia each purchase more than half the grain they consume.

Furthermore, there is considerable intra-regional variation with respect to food security. By our FSI measure it is high in Thailand and South Korea, and very high in Japan. But partly because of lower levels of socio-political stability, most other countries – including China – are classed here as having rather low food security status.

The Far East contains a huge diversity of arrangements linking populations with their food supplies. Today most of the region's people probably buy most of their food in markets and shops. But in certain upland areas of South-east Asia, and the eastern islands of Indonesia, there are still some people living at low densities who are chiefly engaged in shifting cultivation. As in sub-Saharan Africa, natural increase and in-migration are probably causing reductions in the length of fallow in some places. In parts of the Philippines the conditions of agriculture have deteriorated several stages further; population pressures in the lowlands have caused large-scale out-migration, chiefly to the towns, but also to fragile uplands where deforestation and severe land degradation have resulted.[38] The

[38] Broadly analogous processes have occurred in Java, where from early in the twentieth century government 'transmigration' resettlement schemes have moved people to islands of comparatively low population density. In all these locations uncontrolled logging is probably causing more environmental damage than is agriculture, and modern methods of agroforestry – in which the cultivation of trees, crops and animals is integrated – offer some hope of sustainability for poor farmers in the future. See Hung *et al.* (1991) and ICRAF (1993).

broad alluvial deltas of rivers like the Irrawaddy in Myanmar, the Mekong in Cambodia and the Red River in northern Viet Nam provide additional contrast. They contain rural populations living at high densities and largely engaged in wet rice cultivation.

The immensity and density of certain populations in this region have long impressed outsiders. In the *First Essay* Malthus marvelled at China's population:

> When we are assured that China is the most fertile country in the world, that almost all the land is in tillage, and that a great part of it bears two crops every year, and further, that the people live very frugally, we may infer with certainty that the population must be immense.[39]

The quotation hints at an unrivalled degree of agricultural intensification. And the levels of multiple cropping prevailing in China and several of the region's other countries are still exceptionally high.[40] So too are levels of irrigation. Around 1990 half or more of all cropland in China, North Korea, South Korea and Japan was irrigated.[41]

High rural population densities in the past stimulated the development of various ways of conserving soil nutrients which remain important today. They include the close incorporation of nitrogen-fixing legumes into traditional cropping cycles, the use of basin irrigation flood water management in major river deltas, and the application of human and animal excrement as fertilizer (especially in China). The recent huge increase in chemical (overwhelmingly nitrogenous) fertilizer use, particularly in China, can also be viewed as an analogous intensification-response.

The remarkable speed of the modern birth rate declines which have occurred in many countries in the Far East cannot be seen as entirely divorced from these high population densities. In most countries birth rates are already low, or are falling fast. The proportional demographic increases which are forecast for the period 1990–2020 are generally modest – for example, 44 per cent for Indonesia and 29 per cent for China. The UN considers that the size of Japan's population may be decreasing slowly before the year 2010.

[39] See Malthus (1798, p. 56).
[40] See Grigg (1993, pp. 229–31).
[41] See World Resources Institute (1994, p. 294).

However, as we saw in chapter 4, it is as much from the Far East's economic dynamism that its future growth in food demand – and need for increased food imports – is expected to come. The general view which holds today is that most countries on the Pacific rim can anticipate a future of rapidly rising living standards and increasing per capita food consumption. Moreover, especially given the relaxation of restrictions on migration in China, by 2020 perhaps two-thirds of the Far East's people will live in towns.

On this scenario a majority of the region's countries could be importing very significant fractions of their total food consumption by the year 2020. This possibility is underscored by the recent history of sharply rising cereal imports experienced by both South Korea and Malaysia.

To a limited extent this rapid expansion of food demand can be expected to stimulate food production and exports from some poor neighbouring countries like Myanmar and Viet Nam. But the majority of such imports will have to come from outside the region. Our calculations in chapter 4 suggested that by 2020 perhaps 13 to 23 per cent of regional cereal consumption might consist of transfers from outside. This could amount to anywhere between 135 million and 259 million tons. Obviously, China will account for most of this increased demand, possibly followed by Indonesia.

This general prospect seems well-nigh inevitable, at least in the longer run. Already in the mid-1990s many cereal farmers in North America and western Europe are aware of this growing potential market. Thus imports of US wheat into China are generally increasing. And European barley is finding a ready market in the production of Chinese beer. The question is not *whether* the net volumes of cereal (and other food) imports into the Far East will increase.[42] Rather, it is by how much?

This prospect can be portrayed in a daunting light. Lester Brown questions the world's capacity to feed just China. He says the country's rice yields are levelling-off and that wheat yields too are slowing.[43] Given the quantities of cropland being lost to

[42] We refer to the net volumes here because, for example, China exports significant quantities of fruit, vegetables and aquatic products.

[43] See Brown (1994a, p. 189; 1994b). In the early 1990s rice, wheat and coarse grains accounted for 47, 25 and 28 per cent respectively of China's cereal production.

non-agricultural purposes (e.g. factories and housing) and increasing water constraints, Brown considers that *total* grain production in China may fall by at least 20 per cent:

> The resulting grain deficit will be huge – many times that of Japan . . . Allowing only for the projected population increase, with no rise in consumption per person, China's demand for grain would increase to 479 million tons by 2030 . . . [and a] 20 per cent drop in grain output, to 263 million tons, would leave a shortfall of 216 million tons. That level would exceed the world's entire 1993 grain exports.[44]

However, despite these claims, the prospects for China and the rest of the region are probably more complex, less certain and somewhat less disturbing than is being allowed here.

There has actually been very little change in China's area harvested of cereals since the early 1980s – partly due to increased multiple cropping (especially in south China). The same is true throughout the region.[45] It is far from clear that China's irrigation capacity is under any immediate threat.[46] Indeed, even some of the basic facts of China's agriculture are uncertain. For example, there are indications that the country's true levels of cereal production and food stocks could be larger than has hitherto been thought.[47]

According to the calculations in chapter 4 the Far East region will need to achieve an average cereal yield of about 6 tons by 2020. In this context Figure 6.1 shows yield trends for the region's larger countries. In general recent trends have been distinctly upward. This is perhaps particularly true for China – despite the fact that farmers have probably experienced greater incentives to grow non-cereal crops.

It is always hazardous to draw dynamic inferences from comparisons between countries. But it seems plausible to conclude from

[44] See Brown (1994b). See also Brown and Kane (1995, pp. 168–72).

[45] Using FAO data China's areas harvested of cereals in 1980–82 and 1990–92 respectively were 92,905 and 92,380 thousand hectares. Over the same period Indonesia's area increased from 11,721 to 13,706 thousand hectares. Viet Nam, Philippines and Thailand all experienced increases.

[46] See, for example, Nickum (1990).

[47] See USDA (1993b, p. 2), which suggests that because of data inaccuracies China's agricultural situation may need to be reassessed, and that the country may be able to meet its cereal demand from its own production throughout the 1990s. Also statistical sources probably underestimate food output from China's kitchen gardens.

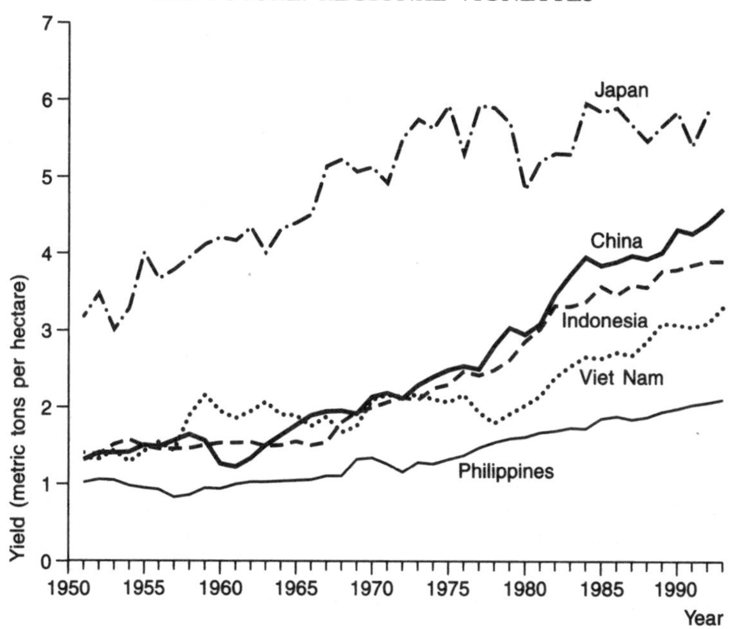

Figure 6.1 Cereal yields for selected countries in the Far East, 1951–93

Figure 6.1 both that most countries in the region have considerable potential to further raise their yields and, crucially, that this also applies to China.

Japan provides one illustration of what can be achieved. Since the mid-1970s its average cereal yield (mainly of irrigated rice) has been about 5.6 tons, although figures over 7 tons are regularly attained by some prefectures. Although Japan imports cereals to use as livestock feed, the country is self-sufficient in rice – the staple cereal food. Indeed, in the 1990s Japanese ricefields have been planted to other crops in order to prevent overproduction.[48] If we assume that China continues to increase its cereal yield at the rate of about 75 kg per hectare per year which it has achieved over the period 1984–93, then the level will reach 5.8 tons by the year 2010, and almost 6.6 tons by 2020.

These linear extrapolations may be on the high side. But there

[48] Prefectures with exceptionally high rice yields are Akita, Aomori and Yamagata. See Gilland (1993, p. 87) and IRRI (1993, pp. 61–3).

are strong grounds to suppose that China can achieve an average cereal yield of over 6 tons by 2020 – mainly by making greater use of already existing improved cereal varieties, raising its chemical fertilizer use (giving greater attention to the mix of nitrogen, phosphate and potash) and by further expanding its irrigation capacity (especially through northward water transference schemes linked to the Three Gorges project on the Yangtze river).[49] Whether Japan's yield represents some kind of 'upper boundary' for China is doubtful.[50] Given China's past achievement in raising yields, the Chinese government's wariness of becoming too dependent upon imports, its very strong commitment to further raise output and the prospect of 'new generation' high yielding cereal varieties, such a conclusion seems very premature. Provided China's political and institutional framework remains sufficiently strong so that it can ensure the necessary volume of investment in agriculture – inter alia by channelling resources from the dynamic east coast industrial sector – then the country should be able to moderate its growing demand for cereal imports.

In summary, the emergence of the Far East, and especially China, as a growing source of future food demand is very probable. But it should not be overstated. In particular, recent cereal yield trends show no obvious signs for alarm. Should there eventually be difficulties in raising China's cereal yields then we can be fairly sure that various responses – such as changes in cropping patterns, greater crop protection, improved plant breeding and the development of intra-regional transfers from countries like Myanmar and Viet Nam – will come into play to dampen the impact on global markets. Certainly, the Chinese government is very wary of the country becoming highly dependent upon imported food. And that is one reason to believe that such a situation will not be allowed to arise.

[49] In 1989–91 49 per cent of China's cropland was irrigated and fertilizer use was 284 kg per hectare of cropland. The corresponding figures for Japan were significantly higher at 62 per cent and 402 kg. On the prospects for greater use of improved seed varieties in China see USDA (1993b). Other likely strategies to increase food output are the upgrading of China's vast grasslands to support more livestock, so relieving pressure on the harvested cereal area, and the switching of more land to potato cultivation.

[50] For this suggestion apropos rice see Brown (1988, pp. 33–4). In fact several countries (e.g. Egypt and North Korea) have higher rice yields than Japan. See IRRI (1993).

LATIN AMERICA

Latin America enjoys a relatively favourable situation with respect to its food supplies and security. And economically and demographically it is comparatively advanced.

We have seen that around 1990 the region was a net cereal importer and that, indeed, it received significant amounts of cereal aid from the United States and Canada. However a focus upon cereals is rather misleading because many Latin American countries – most notably Brazil – are very significant exporters of agricultural products like coffee, cocoa, fruit, meat, soybean and sugar. FAO considers Latin America to be agriculturally the most self-sufficient of the world's developing regions; and theoretical studies indicate that its potential yield levels are high.[51]

As elsewhere, there is tremendous variation within the region. Some rural areas – notably north-east Brazil, and upland parts of Peru and Bolivia – are subject to severe population pressures and very low levels of per capita food availability. In these places many of the expected outcomes have ensued – for example, land degradation and fragmentation, growing landlessness, and out-migration both to the towns and to parts of the Amazon basin in search of cultivable land.[52]

But such demographic pressures are not nearly so prominent if developments in the region are viewed as a whole. For example, deforestation in the Amazon basin owes more to cattle ranchers, logging companies and the Brazilian military than it does to the movement of poor farmers in search of cultivable land.

Governments in Latin America have tended to favour the activities of landowners and commercial companies which control very large tracts of land. These businesses often employ few people, are heavily mechanized and are chiefly oriented towards export crops.[53] As elsewhere, subsistence farmers have generally received little encouragement from national governments which have been primarily concerned with holding urban food prices down.

In many Latin American countries birth rates started falling in the 1960s. Therefore the region's population is projected to grow

[51] See Alexandratos (1988, p. 35) and Plucknett (1993, pp. 9–10).
[52] In Brazil, as in Indonesia, the government has encouraged people to move from densely to sparsely populated rural tracts.
[53] See Heilig (1994, p. 834) and Merrick (1991).

by only 54 per cent over 1990–2020. However, some of the smaller poorer nations – like El Salvador, Bolivia, Guatemala, Haiti and Honduras – are forecast to grow by 70 per cent or more over this period. And in such countries population growth will probably exacerbate problems of aggregate food distribution and availability.

This said, even writers who generally take a despairing view of the future sometimes concede that Latin America's food prospects could be comparatively favourable.[54] The major populous countries of Argentina, Brazil, Chile, Colombia and Mexico all have the agricultural and other resources to meet their projected volumes of food demand. Narrowly the point can be emphasized by noting that Latin America has the second lowest fertilizer use rate of any world region;[55] higher applications will certainly realize significant yield gains. However, this is not to say that the required level of investment in agriculture will be forthcoming. Probably the governments of most countries will continue to ignore distributional issues and the problems of small farmers.

The scenario in chapter 4 implied that by 2020 Latin America could be importing anywhere between 15 and 28 per cent of its cereals. In the early 1990s the figure was about 10 per cent. But this last figure is somewhat perverse since it reflects the recent 'squeezing' of Argentina not only from global markets, but also from long-standing cereal markets within the region (e.g. Brazil). Moreover, these figures make no allowance for the export of non-cereal foods. In the future much will depend upon trends in international prices. Should these prices strengthen sufficiently then it is surely conceivable that Latin America could remain largely self-sufficient in food and even regain its self-sufficiency in cereals. By itself Brazil could probably satisfy a significant fraction of the rising Far East cereal demand.

EUROPE AND THE FORMER SOVIET UNION

Our chief interest in Europe and the former Soviet Union relates to the region's future cereal export potential (though it is also a major importer of other foods). In the early 1990s annual cereal consumption in Europe/FSU was approximately 500 million tons. Although the FSU was the world's largest single cereal importer,

[54] Ehrlich and Ehrlich (1990, p. 80) write that: 'It should be emphasized that, unlike most of Asia, Latin America is not yet pushing against physical or biological limits of agricultural production under available technology.'
[55] See Table 4.4.

taken in conjunction with Europe the region was roughly self-sufficient in grain. All populations in this region are growing comparatively slowly. And over the period 1990–2020 it is expected that the populations of many countries will decrease.[56]

We have seen that people in Europe/FSU are generally well fed. But in eastern Europe, and particularly in parts of the former Soviet Union, future food security could be jeopardized in specific locations by ethnic conflict and warfare. The former Yugoslavia illustrates the danger. But Moldova, Georgia, Chechnya, Armenia, Azerbaijan and Tajikistan are all potential flashpoints.[57]

Around 1990 Europe/FSU was consuming about 58 per cent of its cereals (some 290 million tons) indirectly as livestock feed. In fact there are several ways in which average levels of per capita cereal consumption might decline in the decades ahead.[58] But in chapter 4 it seemed prudent to assume that per capita consumption will remain constant. And on this basis an average cereal yield of about 4.3 tons in 2020 would release about 206 million tons of cereals for export (or aid).

Of course, this projected yield of slightly over 4 tons is simply an extrapolation of recent trends. But some idea of the scope for improvement is given by the sub-regional data presented in Table 6.1. While yields can be expected to improve in all three of the sub-regions shown, it may well be in Europe – rather than the FSU – that the most significant yield gains will come.

The three main food producers of the FSU are Russia, Ukraine and Kazakhstan. In the early 1990s they accounted respectively for 55, 22 and 13 per cent of the FSU's total cereal output. Yields were highest in Ukraine, averaging slightly over 3 tons – nearly double the average yield for Russia and three times that of Kazakhstan.[59] Despite these differentials the potential for future yield increases is probably greatest in Ukraine and parts of western Russia.[60] Grow-

[56] See Appendix.
[57] See Nahaylo (1994).
[58] Since the early 1990s the FSU has dramatically reduced its livestock numbers, and therefore its demand for cereal feed. There is also considerable scope for the introduction of animal breeds which are more efficient in grain-to-meat conversion. It is clear too that socio-political instabilities in the FSU may further reduce the volume of cereal demand. In western Europe there are inklings that demand for meat products may decline in the future. All such developments could release cereals (or cereal cropland) for other uses.
[59] See World Resources Institute (1994, p. 293).
[60] In this and the broader context, it is worth noting Plucknett's observation that: 'A fascinating aspect of yield analysis is that many yield gains do not slow down as yields rise.' See Plucknett (1993, p. 24).

Table 6.1 Average measures of cereal production within sub-regions of Europe/FSU, 1990–92

	Former Soviet Union	*Eastern Europe*	*Rest of Europe*	*Total Europe/ FSU*
Cereal production (million mt)	183.7	91.5	190.0	465.2
Area harvested (million ha)	103.3	25.4	40.1	168.8
Yield (mt per ha)	1.779	3.602	4.738	2.756

ing conditions in Kazakhstan tend to be much more difficult, and are subject to highly erratic rainfall.

Not surprisingly, the massive economic adjustments occurring in the countries of the FSU in the mid-1990s are proving to be extremely difficult. The price of food has shot up, and consumers have switched away from meat towards cheaper products like bread, porridge and potatoes.[61] A significant problem for agricultural reform is that, with the exception of areas close to western Europe, most of these new countries have no real history of private farm ownership. Therefore it is sometimes proving more realistic to try to establish larger collaborative farming companies, rather than individual family farms.[62]

The process of agricultural reform is more advanced in eastern Europe. Hungary and the former Czechoslovakia are both significant cereal producers. In the early 1990s these countries were achieving cereal yields of about 4.8 tons – not far short of the performance of Austria and Denmark where yields averaged about 5.5 tons. Elsewhere in eastern Europe cereal yields are much lower. For example, in Poland – which by itself accounts for about one-third of eastern Europe's cereal output – the average yield in 1990–92 was less than 3 tons. There is widespread agreement that the former communist countries of eastern Europe – especially Poland – have considerable agricultural potential. Given reasonable incentives and agricultural reforms they can certainly become substantial net exporters of food.

[61] See USDA (1993a).

[62] In the short run a very bad harvest in Russia could provoke a modest food crisis, involving international price rises. However, apropos the longer run, it is worth noting that before the First World War Russia was a larger cereal exporter than the United States.

A major influence on the agricultural development of eastern Europe (and perhaps the FSU) is likely to come from improved access to the market of the European Union. A key element here will be reform of the Common Agricultural Policy (CAP) which may well be predicated on an eventual eastwards enlargement of the EU.[63]

The CAP was originally designed when the famines and food shortages associated with the Second World War were still fresh in the minds of European politicians, and when over 40 per cent of people in both France and Italy lived in rural areas. By paying farmers good prices for their crops the CAP became very successful in promoting food production. Because of its redistributive benefits for poorer countries (e.g. Greece and Portugal) and the advantages it has for those richer countries which retain sizeable rural sectors – notably the major agricultural power that is France – the CAP continues to enjoy considerable political support within the EU.[64] And because its amendment needs the agreement of several governments, the policy has proved very difficult to change!

The main reason for CAP reform in the early 1990s has been the agricultural competition which it has aroused with the United States, plus the EU's requirement to agree a GATT trade settlement with the US. The EU's initial response to food overproduction in the 1980s was to pay farmers to set aside land from food production, diversify into non-food crops and promote environmental objectives. However, these strategies have become increasingly expensive. And in the 1990s farmers are facing sharp cuts in the support prices they are paid for their crops – with the biggest reductions occurring for cereals.

The coming years are certainly going to see significant changes in the EU farming sector. There will be further reductions in the agricultural labour force and a move towards bigger farms. Many small farms will close down – leading to increased social welfare costs, and in some countries heightened rural to urban migration. The ageing of Europe's population – including its farmers – is likely to facilitate this process of farm consolidation, as older farmers retire and die.

[63] See NFU (1994, p. 2).

[64] In 1990 approximately 66 per cent of Portugal's population lived in rural areas. The figures for Greece, Italy, Spain and France were respectively 37, 31, 22 and 26 per cent. See United Nations (1991).

In conclusion, there is every reason to expect that Europe/FSU will be able to export 206 million tons of cereals by the year 2020. However, whether it will actually be exporting so much will depend upon factors like international food price movements and, more importantly, the course of its agricultural competition with the US. Such considerations are hard to predict. With increasing international integration, and because of lower land and labour costs, there may well have occurred an eastwards shift in the overall balance of food and agricultural production within the region. The potential for further yield gains is widespread, but is perhaps particularly pronounced in countries such as Poland and Ukraine.

However, cereal yields will probably continue to rise steadily within the countries of the current EU, where despite recent reductions in support prices the underlying increase in yields has remained exceptionally strong.[65] Farmers in western Europe benefit from good soils, favourable climate and high technology. Therefore with steadily increasing efficiency gains (e.g. from farm consolidation) many EU farmers may simply decide to increase their yields and food output – largely irrespective of any changes which are made to the CAP. There will be a transition to lower (but rising) international cereal prices. But, nevertheless, total EU cereal production seems firmly set to continue to rise.[66]

NORTH AMERICA AND OCEANIA

Finally we come to the traditional cereal exporters of the United States, Canada and Australia. It should be obvious that most inhabitants of these countries are very well fed; indeed, many of their people are chronically overfed. Moreover, by any criteria these countries are blessed with exceptionally high levels of national food security.

A fundamental fact informing this enviable situation is that these countries together contain only about 6 per cent of the world's population, but almost 20 per cent of the world's supply of cropland. The dominant agricultural player, of course, is the United States – which accounts for about 80 per cent of the region's total

[65] See Griffin (1993, pp. 34–5) and NFU (1994, p. 23).

[66] See NFU (1994, p. 23). Given upward pressure on international cereal prices in the 1990s partly as a result of the Uruguay Round of GATT, and declines in cereal support prices under the CAP, the 'step down' from EU to world prices confronting many farmers in western Europe will prove to be small.

cereal output. But although neither Canada nor Australia are big cereal producers in comparative international terms, they harvest much larger quantities of cereals than they consume. Therefore both countries are major cereal exporters.[67]

The scenario developed in chapter 4 envisages that together these countries will remain as a principal supplier of cereals for the world market (though being joined – perhaps in a subsidiary role – by Europe/FSU). The scenario requires that over the period 1990–2020 the average cereal yield in North America/Oceania will rise from 3.7 tons to about 5.0 tons – leaving an annual exportable surplus of around 183 million tons.

We have seen that there are no indications of any recent slowdown which might threaten such a future trajectory for cereal yields in this region. On the contrary, what is striking is how readily attainable a yield of 5.0 tons may prove to be – particularly given the potential productivity of US agriculture. Thus although 1992 was admittedly an excellent year (mainly because of an exceptionally good maize harvest in the US) it is worth noting that the average 'regional' cereal yield in that year was already over 4.3 tons.

If there are reasons for concern regarding future cereal yields in this region they relate not to the average levels which may be attained, but rather to a possible rise in yield variability – especially due to increasing frequency of drought in North America.

In comparative international terms the United States has very strong relative advantages (e.g. lower unit costs) in cereal production – particularly vis-à-vis maize and wheat. Canada and Australia have similar strengths with respect to wheat.[68] In time, these advantages may be further enhanced by improved varieties coming from biotechnology.

Faced with the emergence of the (then) EC as a major rival cereal exporting bloc in the early 1980s, the Food Security Act passed by the government of the United States in 1985 provided subsidies to facilitate overseas cereal sales, and set in train a process of reducing cereal support subsidies paid to American farmers.

[67] In 1989–91 average annual net cereal exports (inclusive of cereal aid) from the US, Canada and Australia were about 100 million, 23 million and 14 million tons respectively. These quantities were available for transfer despite the fact that about 70 per cent of cereals consumed in these countries was used as livestock feed.

[68] See Scott and Vollrath (1992, p. 9).

Subsequent US farm legislation has added further impetus towards deregulation, and there have been additional cuts in the cereal support prices paid, especially for wheat and maize. As in Europe/FSU, the cardinal trend is towards greater reliance upon market mechanisms, and a declining role for government agricultural support.

Given that export subsidies in both the EU and US are supposed to fall under the Uruguay Round of the GATT, the United States Department of Agriculture is understandably confident of the ability of US farmers to increase their country's cereal exports and re-establish the US share of the global cereal market.[69] This could probably be achieved without any change in the average areas of US cropland which are currently set aside.[70] But, as we have already implied, it may well be that the United States will actually experience relatively little respite from the export competition coming from the EU, and eventually being joined by producers such as Poland, Russia and Ukraine.

Average farm sizes in the United States, Canada and Australia tend to be very large by any standards. And agriculture in these countries inclines to be very industrial in nature too; 'agribusiness' is a North American term. These characteristics are reflected, for example, in the greater tendency of US farmers to grow industrial crops like rapeseed (for the manufacture of lubricants and plastics) and maize (intended for non-food uses like the production of fuel alcohol and starch). North American farmers are also more advanced than their European counterparts in diversifying towards off-farm income. And the United States is noteworthy for its Conservation Reserve Program, which promotes tree planting, soil conservation, wetland preservation and the improvement of water quality.[71] Nevertheless, despite such developments in the agricultural sector, the United States, Canada and Australia have all experienced a continued population drift away from farming. The prospect must be that this trend will continue. And even in the United States the coming decades may see market forces and

[69] See the USDA quotation cited in NFU (1994, Appendix II, p. ix).
[70] Estimates of areas of cropland being idled should be viewed with caution, since they fluctuate from year to year and are heavily dependent upon the particular definitional criteria which are employed. See, for example, USDA (1992, p. 8).
[71] See, for example, USDA (1992, 1993c). For similar reasons farmers in the EU also face a future in which their livelihoods will become increasingly dependent upon financial support tied to meeting environmental objectives.

technological innovations leading to still larger farm sizes, with fewer farmers.

Finally, a word is in order regarding the probable implications of the North American Free Trade Agreement (NAFTA) between the US, Canada and Mexico. This represents yet another aspect of an increasingly interdependent and specialized agricultural world. Producers of fruit and vegetables in the United States, and especially Canada, are likely to face much stiffer competition from Mexican growers. Conversely, farmers in Mexico are unlikely to be able to compete in the production of maize, which presently is their principal cereal crop. So Mexico seems set to become increasingly reliant upon its northern neighbours for both cereals and meat.[72] Over the longer run the influence of NAFTA will probably become stronger elsewhere in Latin America too.

In conclusion, the agricultural production and trading systems of the United States, Canada and Australia are extremely advanced. But, ironically, some of the problems currently facing farmers in these countries have been complicated by their own success in raising farm productivity. It is hard to envisage that these countries will not achieve an average cereal yield of 5 tons by the year 2020. The farmers of North America/Oceania will probably have the upper hand; but, almost certainly, they face a future of extended competition with their counterparts in Europe, and eventually with farmers in parts of the FSU. This competition will exert distinct downward pressure on international prices, although clearly this could be partly offset by growth in demand. Nevertheless, even in North America and Australia there must be a strong expectation that the volume of agricultural employment will continue to decline in the decades ahead.

CONCLUDING REMARKS

The regional vignettes presented here broadly substantiate the scenario developed in chapter 4. To some degree this was to be expected – since one aim of this chapter was to briefly summarize what has already been concluded for each world region.

However, another aim was to address linkages between demographic and agricultural conditions within each region. And, while

[72] The situation is broadly analogous with that facing the EU with respect to Turkey, Egypt and Morocco.

there were certainly sometimes differences of detail and emphasis, it can be claimed with all sincerity that in each case the thrust of the regional literatures and opinions surveyed for this chapter was in general agreement with the sketches presented above. In that sense our world regional scenario (from chapter 4) gains a measure of independent support from arguments developed here closer to the ground.

Inevitably, views regarding future prospects are heavily coloured by recent developments. However, it is very difficult to envisage that the currently prevailing broad process of global market integration will somehow be reversed during the next few decades. A continuing move towards agricultural trade liberalization and specialization seems well-nigh inevitable (and, of course, extends far beyond agriculture in its ramifications). Nevertheless, it is noteworthy that this general trend may have analogous consequences for people living in widely divergent settings – for example, contributing to farm consolidation and a drift from the land in many locations.

At the regional level, however, it is likely that currently prevailing views will be less dependable guides as to the future. Here two remarks seem appropriate. First, perhaps China's economy will not continue to expand at quite the giddy speed which is now so widely supposed. And second, we can only hope that the generally despondent consensus regarding sub-Saharan Africa's long-term prospects turns out to be too heavily influenced by the very unfortunate food production trends of the recent past.

7

CONCLUSIONS, FORECASTS, CAVEATS, TEMPERED HOPE

PREAMBLE

Food is a topic full of opposites. Nearly everyone has views. There can be times of feast, and times of famine. Some see human hunger as primarily a problem of production, while others emphasize issues of distribution. There are those who advocate national self-sufficiency in food, and those who favour policies of unrestricted free trade. Many people regard low food prices as a boon, but most farmers want higher prices. Usually there is some merit in both opposing views. Usually reality and sense rest somewhere in between.

And then there are pessimists and optimists on the issue of whether food output will keep up with demographic growth. This book has examined recent trends in the relationship between population growth and food production, and it has assessed global prospects for the period to 2020. In general, for this time horizon, we find the pessimistic neo-Malthusian view seriously wanting; several of its fundamental claims are demonstrably unsound.

Alas, however, this does not mean that everything is rosy. It is obvious that much is unacceptable about the current world food situation. There are important elements of warning in the pessimistic position.

CONCLUSIONS

Hunger and food insecurity are widespread in the modern world. Sub-Saharan Africa suffers chronically from both phenomena. But South Asia probably contains the greatest number of hungry people. Levels of per capita food availability are somewhat higher in the Far

East, yet there is still much hunger there, and levels of national food security in the region are generally low. In the Middle East relatively high levels of average food availability often belie the extent of food insecurity – and this region is already singularly dependent upon imported food. Even Latin America – with its relatively rich agricultural resource base – confronts serious food problems.

However, while population growth certainly sometimes adds to the task of raising levels of per capita food production (and consumption), it is not the principal cause of contemporary world hunger. It is important to remember that, until relatively recently in human history, 'hunger' (at least as we would define it today) has probably been the lot of most people, in most places, at most times.

Moreover, we have seen that in recent years and decades food production has generally grown faster than population. It is true that world grain output has fallen behind demographic growth since 1984. But the reason for this lies largely in the rich world's overproduction of cereals, compared to the quantities which could reasonably be sold (even at heavily subsidized prices) or otherwise dispensed. Very low selling prices discouraged many of the world's farmers from growing grain. Deliberate curbs on cereal production – especially, but not exclusively, in North America – constitute a large part of the explanation for the fall in global per capita cereal output since the early 1980s.

Only in sub-Saharan Africa do we find credible evidence that cereal (and to a lesser extent food) output has failed to keep abreast of population growth, partly because of the speed and volume of that growth itself.

But in all other world regions food production has kept ahead of population increase. Since the early 1980s farmers have often switched out of cereals towards more remunerative food (and non-food) crops. Diets have generally become more diverse. Recent progress in raising per capita food production has been – and continues to be – greatest in the world's two most populous regions, i.e. the Far East and South Asia. Moreover, the expansion of global trade in food means that inter-regional transfers now play an increasing role in moderating changes in per capita food consumption. Even sub-Saharan Africa has had its dismal recent food production performance significantly offset by increased food transfers from outside.

Underpinning this general progress has been a steady rise in food crop yields. At the aggregate world level there is no reason to be particularly alarmed by recent trends (though, clearly, things would be better if yield increments were greater). Importantly, we have seen that yields tend to increase 'arithmetically' rather than 'geometrically'. In most regions recent increments in average cereal yields per hectare compare favourably with previous experience. It is equally important to appreciate that raising yields is an immensely complicated and multifaceted process. It is folly to put one's hopes in wonder breakthroughs; it is folly to conceptualize future food production prospects largely in such terms. At the aggregate level, even the effects of 'green revolutions' (an oversimple expression) become rather subsumed in yield trends.

FORECASTS

So, several pivotal neo-Malthusian assertions regarding the relationship between population and food during the recent past can be firmly rejected. In turn, this makes it easier for us to construct a more tolerable – if far from perfect – vision of the future. The analysis in this book broadly supports the following propositions.

The average level of world per capita food production (and consumption) in the year 2020 will probably be fairly similar to the level which applied around 1990. However, as we have repeatedly stressed, this will be entirely compatible with *rises* in per capita food consumption in most regions. Virtually all of the increase in world population in the period to 2020 will occur in regions with lower levels of per capita food consumption. Other things being equal, over time this will tend to bias downwards any simple (and therefore rather misleading) measure of global per capita food availability.

As we have seen, demographic growth has become the increasingly predominant cause of the expansion of world food production. This trend will continue. Using cereals as a rough proxy for food in general, it is reasonable to expect that between 80 and 90 per cent of the rise in world food output over the period to 2020 will be due to increased demand generated by population growth.

Our judgement is that general demographic, socio-economic and political conditions in sub-Saharan Africa may be so difficult that there will be little change – perhaps even some deterioration – in average levels of per capita food consumption there. But in the remaining predominantly developing regions – perhaps especially in the Far East – the scenario developed in chapter 4 envisages modest gains in per capita consumption. Assuming that levels of consumption in the two developed world regions remain constant, then nearly 3 billion tons of cereals will need to be harvested in 2020 in order to satisfy global demand.

Given no unforeseen huge calamity, the world's farmers will certainly be able to meet this volume of demand. To do so the average global cereal yield will have to reach about 4 tons per hectare – a feasible target on the assumption of a continuing linear trend. There may also have to be a modest expansion of the world's harvested cereal area – through some mixture of increased multiple cropping, reduced set aside and cultivation of new land.

World agriculture will certainly see a steady rise in fertilizer use. By 2020 global applications of synthetic nitrogen will probably have doubled. The world's irrigated area will also continue to expand – especially in the Far East and South Asia. Global irrigation capacity in 2020 will still be well above 40 hectares per thousand population. And, of course, there will be further new varieties of food crops with higher yields.

In the future, as in the past, a multitude of other factors will interact to raise world agricultural productivity. But, in particular, we can expect that there will be greater and greater emphasis upon information-intensive farm management procedures. The decisions of farmers across a whole range of issues – e.g. water use, soil quality, nitrogen up-take – will become increasingly tailored to the requirements of individual fields. For example, sometimes this will involve detailed study of high-resolution satellite photographs; other times farmers will be better interpreting subtle differences and changes in the colours of their foodcrop leaves.

Returning to the regional level of analysis, it appears inevitable that there will be a growing future *mismatch* between the expansion of food demand and food supply. Inevitably, therefore, the international trade in food will greatly increase. The illustrative calculations in chapter 4 implied that the volume of inter-regional cereal transfers around 2020 could well be three times greater than was the case around 1990.

The world seems set to have two major food exporting blocs (recall that for most of the period since the Second World War there has been only one). The recent rivalry between the traditional cereal exporters of North America and the EU will continue (being eventually joined and complicated by countries in eastern Europe and the FSU). Notwithstanding GATT agreements and the existence of the World Trade Organization, we can confidently expect many more agricultural trade wrangles between these two exporting blocs in the coming years! A major, long-term rise in world food prices might ease this rivalry. The issue of future price movements is hard to fathom. But the last hundred years of generally falling international grain prices suggest that such a sustained price rise is unlikely to occur.

Turning to the food importing regions, with the possible exception of Latin America, they will all be significantly more dependent upon imports to meet their 2020 food requirements (in both absolute and proportional terms) than was the case in 1990. The Middle East could be importing half its total cereal supplies. The volume of demand stemming from the Far East (particularly China) will certainly be very large – amounting to perhaps 13 to 23 per cent of its total consumption. But it seems likely that South Asia will be much less reliant upon imports than the Far East. It is our expectation that most countries in most regions will be able to finance most of their imports. But this may not be true for sub-Saharan Africa – which by 2020 may need to import a quarter or more of its total cereal supplies simply to sustain its current levels of per capita consumption.

It is worth remarking that this broad forecast envisages that roughly two-thirds of the increase in global cereal production in the period to 2020 will occur in the predominantly developing, food importing, world regions themselves. Thus, for example, when people ask 'Who will feed China?', the answer is plain: 'Mostly, the Chinese'. But so far as export capacity is concerned, it is obvious that the two most developed regions (essentially North America and Europe) enjoy the best of any prevailing 'comparative advantage' – given their supplies of cropland, their agricultural systems and their associated industries.

CAVEATS

Inevitably, there are significant provisos and unknowns.

The first is sub-Saharan Africa. Extremely rapid demographic growth is just one of this region's serious difficulties. It seems unlikely that food production can be raised as fast as the projected population growth. Therefore the volume of food transfers into the region must increase. The issue then arises as to the basis, and extent, of these future transfers. Some sub-Saharan countries will be able to finance imports. But it seems probable that other countries will become more dependent upon supplies of food aid provided mainly for humanitarian reasons. Perhaps bargains will be struck whereby environmental conservation (and other) programmes are instigated partly in exchange for food. But there must be a chance that food transfers into the region will not expand at the envisaged rate, and, therefore, that sub-Saharan Africa's already meagre levels of per capita food consumption will decline.

Then there is the issue of the total volume of human hunger in 2020, i.e. the number of undernourished people there will be. As we have already noted, the foregoing broad forecast anticipates that average levels of food availability will remain approximately constant in sub-Saharan Africa and the two most developed world regions, but that there will be modest rises in per capita consumption in the Middle East, South Asia, Latin America and the Far East. However, without addressing the knotty issue of future trends in income (and therefore food) distribution, it is impossible to say whether there will be fewer (or more) undernourished people alive in 2020 than was the case in 1990. Our discussion of institutional structures in chapter 5 does not inspire confidence that the number will be very much less. Perversely, much current economic 'wisdom' appears to find favour in greater income inequality.

Looking at sub-Saharan Africa, the Far East and South Asia, there are already signs that many poor people are worse off in terms of their food availability and food security – precisely because of recent measures of economic 'liberalization'. One problem with evaluating food issues is that many different criteria can be employed. Thus average levels of per capita food consumption can increase simultaneously with the numbers of hungry people. Or to give another example, in terms of its future import requirements South Asia's prospects may appear to be significantly better than those of the Far East – because the former region will probably import considerably less food. But these 'better' prospects will partly result from much lower average levels of future food consumption.

Another important proviso relates to variability. Chapter 3 presented fairly strong evidence that harvest variability in both sub-Saharan Africa and North America/Oceania is on the rise. These are the two world regions which are most dependent upon rainfed agriculture. And in both cases the basic explanation for the rising trend is more frequent drought. These disturbing developments may continue; or, at least, they may not retreat. For sub-Saharan Africa this could constitute yet another deteriorating circumstance. The most significant effect of more drought in North America/ Oceania may operate through greater international grain price volatility – in a world in which many poor countries are increasingly dependent upon cereal imports.

This said, it should be recalled that in most world regions there are no signs of a trend towards greater harvest variability (despite the fact that it is sometimes contended that rising yields are inherently less stable). This point of reassurance applies to the world's second potential major food exporting bloc of Europe/ FSU. So one's views as to how much of a problem heightened harvest variability may be in the next few decades must be mixed. We certainly cannot entirely preclude a repeat of the ENSO-inspired droughts which occurred in 1972–74. But humanity's capacity to deal with any future crisis should also be somewhat improved.

Unless the global climate really does go awry during the period to 2020, in most world regions it is hard to see drought, by itself, causing famine – as it virtually did, for example, in several parts of Asia around 1972–74. But, again, sub-Saharan Africa could be a different story. In any case, for many countries in this region we assess overall levels of national food security (including a socio-political component) to be so low as to virtually ensure the continuation of periodic food crises in sub-Saharan Africa.

For several reasons in this book we have deliberately looked forward just to the year 2020. One advantage of this is that we are dealing (now) with a future period of about 25 years. For the writer, and some readers, it is fairly easy – in one's 'mind's eye' – to look backward for the same length of time, i.e. to around 1970. The coming quarter century will certainly have its surprises (perhaps the best would be if circumstances in sub-Saharan Africa genuinely improved). But on the issue of the relationship between world population growth and food supplies it is hard to envisage that the situation will get very much worse in the way that some neo-Malthusians evidently suppose.

However, wondering beyond 2020, one must express some reservations about what could eventually happen to world food production if there is sustained global warming. Likewise, recall that in chapter 4 we could not completely rule out the possibility that neo-Malthusian feedbacks might eventually operate, to some extent, in certain locations and perhaps particularly over the longer run.

TEMPERED HOPE

Despite these caveats, there is fair reason to expect that in the year 2020 world agriculture will be feeding the larger global population no worse – and probably a little better – than it manages to do today. This all adds up to tempered hope.

Obviously the task won't be easy (it has not been easy in the past). For most poor countries we can firmly reject any optimistic notion that their demographic growth alone will somehow spontaneously generate commensurate increases in food production. Population growth may be a slow-acting process, the independent effects of which are sometimes hard to isolate in statistical analyses. But there can be no doubt at all that contemporary world population growth *is* making the task of satisfactorily feeding humanity significantly harder to accomplish than it would otherwise be. And population growth is going to be the paramount cause of increased food production in the coming decades – a fact with countless ramifications.

On the other hand, we have also seen that the modern neo-Malthusian case suffers from major questions of credibility. Pronouncements from this quarter are often deceptively simple and have to be read with very great care. There can be problems of selectivity and representation. Moreover, these problems of credibility are not confined to a neo-Malthusian core, but are sometimes displayed by representatives of mainstream agencies concerned with world food issues.

Simple claims and soundbites are doubtless easy to convey. And in most years it is usually possible to represent some of 'last year's' statistics in a disturbing light. But the relationship between population and food is complex. Indeed, one major reason for tempered hope regarding humanity's prospects to 2020 derives precisely from this very complexity. A host of factors intervene between population and food to provide a measure of flexibility and adaptability. This book will inevitably age in its details. But, obviously, we believe that its essential propositions will gain strength with the passage of time.

In a sense, we come full circle to Malthus. In chapter 1 we noted that his *considered* views about the relationship between population and food – beginning with the publication of the *Second Essay* in 1803 – differ in major respects from what has come to be associated with his name. In particular, the mature scholar that was Thomas Robert Malthus attached great import to the careful collection and reading of empirical data. We could do with more of this quality today. And, perhaps more than anyone else, he was certainly aware of the many complexities linking populations with their supplies of food.

APPENDIX: COUNTRY DATA AND FOOD SECURITY INDEX, AROUND 1990

Region/country	*Population (million)*	*Projected population change 1990–2020 (%)*	*Per capita daily calorie supply*	*GNP per capita $US*	*Cereal output as % of consumption*	*Index of socio-political stability*	*Food security index (FSI)*	*FSI rank Australia = 1 Angola = 91*
Sub-Saharan Africa								
1 South Africa	37.1	79	3133	2530	120	11	0.54	20
2 Madagascar	12.6	143	2156	230	94	10	−0.18	38
3 Mozambique	14.2	121	1805	80	35	8	−0.84	89
4 Zimbabwe	9.9	84	2256	640	135	6	−0.11	33
5 Angola	9.2	155	1807	620	47	2	−1.11	91
6 Zambia	8.2	112	2016	420	85	9	−0.35	55
7 Malawi	9.4	112	2049	200	82	11	−0.24	44
8 Zaïre	37.4	145	2130	220	75	3	−0.75	83
9 Tanzania	25.6	120	2195	110	98	4	−0.51	69
10 Burundi	5.5	120	1948	210	93	3	−0.70	81
11 Rwanda	7.0	106	1913	310	93	2	−0.78	85
12 Uganda	17.9	136	2178	220	99	6	−0.39	58
13 Kenya	23.6	140	2064	370	90	4	−0.61	74
14 Somalia	8.7	118	1874	120	54	2	−1.06	90
15 Ethiopia	47.4	136	1667	120	84	5	−0.76	84
16 Cameroon	11.5	126	2208	960	78	5	−0.54	72
17 Chad	5.6	110	1743	190	92	5	−0.68	78
18 Nigeria	96.2	123	2200	290	70	3	−0.74	82
19 Niger	7.7	154	2239	310	93	8	−0.27	48
20 Ghana	15.0	127	2144	390	74	7	−0.49	65
21 Ivory Coast	12.0	165	2568	750	68	5	−0.46	64
22 Guinea	5.8	133	2242	440	80	5	−0.54	71
23 Burkina Faso	9.0	114	2219	330	92	7	−0.35	54
24 Mali	9.2	137	2259	270	93	10	−0.14	35
25 Senegal	7.3	109	2322	710	55	7	−0.53	70
Middle East								
26 Morocco	24.3	58	3031	950	76	6	−0.15	37
27 Algeria	24.9	72	2944	2060	31	2	−0.68	79
28 Tunisia	8.1	56	3122	1440	55	5	−0.29	50
29 Egypt	56.3	63	3310	600	59	4	−0.27	49
30 Sudan	24.6	114	2043	540	79	2	−0.80	86
31 Turkey	56.1	54	3196	1630	98	6	0.09	29
32 Syria	12.3	146	3122	1000	67	2	−0.41	61
33 Saudi Arabia	16.0	136	2929	7050	55	2	−0.36	57
34 Yemen	11.3	165	2300	610	33	5	−0.81	87
35 Iraq	18.1	115	2887	1940	44	2	−0.63	75
South Asia								
36 Iran	58.9	96	3181	2490	71	3	−0.25	45
37 Afghanistan	15.0	174	2280	400	85	2	−0.67	77
38 Pakistan	121.9	115	2280	380	94	8	−0.24	43
39 India	850.6	56	2229	350	100	8	−0.23	41
40 Nepal	19.3	96	2205	170	100	9	−0.18	39
41 Bangladesh	108.1	71	2037	210	90	10	−0.25	46
42 Sri Lanka	17.2	40	2246	470	65	7	−0.50	67
Far East								
43 China	1155.3	29	2641	370	96	2	−0.45	63
44 Myanmar (Burma)	41.8	71	2454	210	101	2	−0.50	66
45 Thailand	55.6	29	2280	1420	134	8	0.05	32
46 Cambodia	8.8	104	2166	400	98	7	−0.32	53
47 Viet Nam	66.7	67	2233	215	106	2	−0.56	73
48 North Korea	21.8	47	2843	1240	93	2	−0.35	56
49 South Korea	42.9	24	2826	5400	45	12	0.09	30
50 Japan	123.5	0	2921	25,430	35	12	0.74	17
51 Philippines	60.8	63	2341	730	86	9	−0.19	40
52 Malaysia	17.9	66	2671	2320	43	7	−0.40	60
53 Indonesia	182.8	44	2605	570	96	3	−0.40	59

Region/country	*Population (million)*	*Projected population change 1990–2020 (%)*	*Per capita daily calorie supply*	*GNP per capita $US*	*Cereal output as % of consumption*	*Index of socio-political stability*	*Food security index (FSI)*	*FSI rank Australia = 1 Angola = 91*
Latin America								
54 Mexico	84.5	55	3062	2490	78	8	0.06	31
55 Cuba	10.6	17	3129	2000	18	2	−0.69	80
56 Haiti	6.5	84	2005	370	42	6	−0.82	88
57 Dominican Republic	7.1	51	2310	830	39	9	−0.51	68
58 Guatemala	9.2	114	2254	900	72	7	−0.44	62
59 El Salvador	5.2	76	2317	1110	67	10	−0.26	47
60 Honduras	5.1	92	2210	590	68	10	−0.31	51
61 Venezuela	19.5	69	2443	2560	55	10	−0.23	42
62 Colombia	32.3	47	2453	1260	84	9	−0.14	36
63 Ecuador	10.3	65	2399	980	74	11	−0.12	34
64 Peru	21.6	61	2037	1160	51	7	−0.66	76
65 Brazil	148.5	49	2730	2680	91	10	0.13	26
66 Bolivia	6.6	86	2013	630	71	11	−0.31	52
67 Chile	13.2	44	2484	1940	91	12	0.12	27
68 Argentina	32.5	36	3068	2370	177	11	0.88	15
Europe/FSU								
69 Greece	10.2	−2	3775	5990	116	12	0.97	13
70 Bulgaria	9.0	−12	3695	2250	94	12	0.67	19
71 Yugoslavia (fmr)	23.8	9	3545	3060	101	6	0.31	23
72 Romania	23.2	−5	3081	3445	95	9	0.27	25
73 Hungary	10.4	−9	3608	2780	113	13	0.83	16
74 Czechoslovakia (fmr)	15.7	13	3574	3140	99	3	0.12	28
75 Austria	7.7	7	3486	19,060	120	14	1.44	8
76 Switzerland	6.8	14	3508	32,680	73	14	1.61	6
77 Italy	57.0	−6	3498	16,830	81	13	1.05	11
78 Spain	39.3	−2	3472	11,020	95	13	0.94	14
79 Portugal	9.9	−1	3342	4900	51	14	0.45	22
80 France	56.7	7	3593	19,490	198	13	1.94	2
81 Belgium	10.0	4	3925	15,540	46	14	1.03	12
82 Netherlands	15.0	9	3078	17,320	39	14	0.69	18
83 UK	57.4	6	3270	16,100	120	13	1.18	10
84 Germany	79.4	−2	3523	22,320	104	13	1.40	9
85 Poland	38.1	8	3426	1690	93	12	0.53	21
86 Denmark	5.1	−1	3639	22,080	142	14	1.75	5
87 Sweden	8.6	12	2978	23,660	132	14	1.46	7
88 Former Soviet Union	281.3	20	3380	9211	84	5	0.27	24
N. America/Oceania								
89 Canada	27.8	33	3242	20,470	178	14	1.76	4
90 USA	249.9	28	3642	21,790	146	14	1.77	3
91 Australia	16.9	40	3302	17,000	292	14	2.41	1

Notes: (a) Virtually all of the measures given above relate to years in the period 1988–94 inclusive. The principal sources of data used are listed below. But in respect of each measure it was occasionally necessary to supplement these sources (usually with data for an earlier year or period) due to reasons of unavailability, or changes in national jurisdictions (e.g. in the cases of the former Yugoslavia, Czechoslovakia and Soviet Union)

(b) As noted in chapter 2, the Freedom House measure of liberty has been used as a rough proxy for socio-political stability. Here it ranges from 14 = freest to 2 = least free. We leave it to others to further develop the integration of measures of socio-political stability into indices of national food security

(c) The crude and approximate nature of the FSI values should be stressed. In discussing food security for individual countries many qualifications would need to be made. The rough nature of the FSI values is reflected, for example, in the fact that they are based on individual measures which refer to slightly different periods and years. The values are probably best considered as referring to circumstances in the early 1990s

Principal sources: FAO (1993), Freedom House (1995), United Nations (1994), World Bank (1992), World Resources Institute (1992)

BIBLIOGRAPHY

African National Congress (1994) *The Reconstruction and Development Programme: A Policy Framework*, Johannesburg: Umanyano Publications.

Agrawal, A.N., Varma, H.O. and Gupta, R.C. (1993) *India Economic Information Yearbook 1992–93*, New Delhi: National Publishing House.

Alamgir, M. (1980) *Famine in South Asia: Political Economy of Mass Starvation*, Cambridge, Mass.: Oelgeschlager, Gunn and Hain.

Alamgir, M. and Arora, P. (1991) *Providing Food Security for All*, London: Intermediate Technology Publications.

Alexandratos, N. (1988) *World Agriculture: Toward 2000, An FAO Study*, London: Belhaven Press.

Allan, T. (1995) 'The political economy of Jordan catchment water', in J.A. Allan and J.H.O. Court (eds) *Water in the Jordan Catchment Countries*, SOAS Water Issues Group, London: School of Oriental and African Studies.

Anderson, K. and Tyers, R. (1991) *Global Effects of Liberalizing Trade in Farm Products*, London: Harvester-Wheatsheaf.

Anzala, A.O., Ndiny-Achola, J.O., Ngugi, E.N., Simonsen, J.N., Kreiss, J.K. and Plummer, F.A. (1993) 'Rapid progression to disease in African prostitutes with HIV-1 infection', paper presented at The First National Conference on Human Retroviruses and Related Infections, co-sponsored by National Foundation for Infectious Diseases and the American Society for Microbiology, Washington, DC, 12–16 December.

Banister, J. (1987) *China's Changing Population*, Stanford, Calif.: Stanford University Press.

Barnett, V., Payne, R. and Steiner, R. (eds) (1995) *Agricultural Sustainability: Economic, Environmental and Statistical Considerations*, Chichester: John Wiley.

Basu, D.R. (1984) 'Food policy and the analysis of famine', *Indian Journal of Economics* 64, 254: 289–301.

Bennett, M.K. (1949/1992) 'Population and food supply: The current scare', in *Scientific Monthly* 68, 1, reprinted in *Population and Development Review* 18, 2: 341–58.

Bhat, P.N., Preston, S. and Dyson, T. (1984) *Vital Rates in India, 1961–81*,

Committee on Population and Demography, Report 24, Washington, DC: National Academy Press.

Bilsborrow, R.E. (1987) 'Population pressures and agricultural development in developing countries: A conceptual framework and recent evidence', *World Development* 15, 2: 183–203.

Blaxter, K. (1986) *People, Food and Resources*, Cambridge: Cambridge University Press.

Bloomfield, P. (1992) 'Trends in global temperature', *Climatic Change* 21, 1: 1–16.

Boserup, E. (1965) *The Conditions of Agricultural Growth*, London: George Allen and Unwin.

——— (1990) *Economic and Demographic Relationships in Development*, Baltimore, Md: Johns Hopkins University Press.

Braun, J. von (1991) 'A policy agenda for famine prevention in Africa', Food Policy Report, Washington, DC: International Food Policy Research Institute.

Braun, J. von and Puetz, D. (eds) (1993) *Data Needs for Food Policy in Developing Countries: New Directions for Household Surveys*, Washington, DC: International Food Policy Research Institute.

Braun, J. von, Bouis, H., Kumar, S. and Pandya-Lorch, R. (1992) 'Improving food security of the poor: Concept, policy, and programs', Washington, DC: International Food Policy Research Institute.

Braun, J. von, Hopkins, R.F., Puetz, D. and Pandya-Lorch, R. (1993) 'Aid to agriculture: Reversing the decline', Food Policy Statement 17, Washington, DC: International Food Policy Research Institute.

Brown, L.R. (1988) 'The changing world food prospect: The nineties and beyond', *Worldwatch Paper 85*, Washington, DC: Worldwatch Institute.

——— (1991) 'The new world order', in L.R. Brown, C. Flavin and S. Postel (eds) *State of the World 1991*, London: Earthscan Publications Ltd.

——— (1993) 'A new era unfolds', in L.R. Brown, C. Flavin and S. Postel (eds) *State of the World 1993*, New York: W.W. Norton.

——— (1994a) 'Facing food insecurity', in L.R. Brown, C. Flavin and S. Postel (eds) *State of the World 1994*, London: Earthscan Publications Ltd.

——— (1994b) 'Question for 2030: Who will be able to feed China?', *International Herald Tribune*, Paris, 28 September.

——— (1994c) 'Fertilizer use keeps dropping', in L.R. Brown, H. Kane and D.M. Roodman (eds) *Vital Signs 1994, 1995*, London: Earthscan Publications Ltd.

Brown, L.R. and Eckholm, E.P. (1974) *By Bread Alone*, New York: Praeger.

Brown, L.R. and Kane, H. (1995) *Full House: Reassessing the Earth's Population Carrying Capacity*, London: Earthscan Publications Ltd.

Brown, L.R., Kane, H. and Ayres, E. (eds) (1993) *Vital Signs 1993*, New York: W.W. Norton.

Brown, L.R., Kane, H. and Roodman, D.M. (eds) (1994) *Vital Signs 1994, 1995*, London: Earthscan Publications Ltd.

Caldwell, J.C. (1975) 'The Sahelian drought and its demographic implications', Paper 8, Overseas Liaison Committee, Washington, DC: American Council on Education.

Caldwell, J.C. and Caldwell, P. (1993a) 'The South African fertility decline', *Population and Development Review* 19, 2: 225–62.

——— (1993b) 'The nature and limits of the sub-Saharan African AIDS epidemic: Evidence from geographic and other patterns', *Population and Development Review* 19, 4: 817–48.

Carruthers, I. (1993) 'Going, going, gone! Tropical agriculture as we knew it', Occasional Paper 93/3, Wye College, University of London.

Cassman, K.G., Kropff, M.J., Gaunt, J. and Peng, S. (1993) 'Nitrogen use efficiency of rice reconsidered: What are the key constraints?', *Plant and Soil* 155/156: 359–62.

CGIAR (1992) *CGIAR Annual Report 1992*, Washington, DC: CGIAR Secretariat.

Coale, A.J. (1984) *Rapid Population Change in China, 1952–1982*, Committee on Population and Demography, Report 27, Washington, DC: National Academy Press.

Conway, G., Lele, U., Peacock, J. and Piñeiro, M. (1994) *Sustainable Agriculture for a Food Secure World: A Vision for International Agricultural Research*, Washington, DC: CGIAR Secretariat.

Crosson, P. and Anderson, J.R. (1992) 'Resources and global food prospects', World Bank Technical Paper 184, Washington, DC: The World Bank.

Daily, G.C. and Ehrlich, P.R. (1990) 'An exploratory model of the impact of rapid climate change on the world food situation', *Proceedings of the Royal Society of London, B* 241: 232–44.

Demeny, P. (1994) 'Population and development', IUSSP Distinguished Lecture Series at the United Nations International Conference on Population and Development, Cairo, 1994, published in Liège by the International Union for the Scientific Study of Population.

de Waal, A. (1993) 'War and famine in Africa', in J. Swift (ed.) 'New approaches to famine', *IDS Bulletin* 24, 4: 33–40.

Downing, T.E. (1992) 'Climate change and vulnerable places: Global food security and country studies in Zimbabwe, Kenya, Senegal and Chile', Research Report 1, Environmental Change Unit, University of Oxford.

Downing, T.E. and Parry, M. (1994) 'Climate change and population supporting capacity: Potential impacts and vulnerable regions', in B. Zaba and J. Clarke (eds) *Environment and Population Change*, Liège: Derouaux Ordina Editions.

Duddin, M. and Hendrie, A. (1992) *World Land and Water Resources*, London: Hodder and Stoughton.

Dyson, T. (1991) 'On the demography of South Asian famines, Part II', *Population Studies* 45, 2: 279–97.

——— (1993) 'Population and food in South Asia: Recent trends and prospects', paper presented at the XVII Annual Conference of the Indian Association for the Study of Population, held at Annamalai University, Tamil Nadu, 16–19 December.

——— (1994a) 'Population growth and food production: Recent global and regional trends', *Population and Development Review* 20, 2: 397–411.

Dyson, T. (1994b) 'World population growth and food supplies', *International Social Science Journal* 141: 361–85.

Dyson, T. and Maharatna, A. (1992) 'Bihar famine, 1966–67, and Maharashtra drought, 1970–73: The demographic consequences', *Economic and Political Weekly* 27, 26: 1325–32.

Ehrlich, P.R. (1968) *The Population Bomb*, New York: Ballantine Books.

Ehrlich, P.R. and Ehrlich, A.H. (1990) *The Population Explosion*, New York: Simon B. Schuster.

Ehrlich, P.R., Ehrlich, A.H. and Daily, G.C. (1993) 'Food security, population, and the environment', *Population and Development Review* 19, 1: 1–32.

Falkenmark, M. (1991) 'Rapid population growth and water scarcity: The predicament of tomorrow's Africa', in K. Davis and M.S. Bernstam (eds) *Resources, Environment, and Population*, Supplement to *Population and Development Review* 16, New York: Oxford University Press.

FAO (1987a) *World Crop and Livestock Statistics 1948–85*, Rome: Food and Agriculture Organization.

——— (1987b) *The Fifth World Food Survey*, Rome: Food and Agriculture Organization.

——— (1992) 'World food supplies and prevalence of chronic undernutrition in developing regions as assessed in 1992', Document ESS/MISC/1/92, Rome: Food and Agriculture Organization.

——— (1993) *Production Yearbook, 1992* 46, Rome: Food and Agriculture Organization.

——— (1994) *Production Yearbook, 1993* 47, Rome: Food and Agriculture Organization.

FAO/WHO (1992) *Nutrition: The Global Challenge*, Rome: Food and Agriculture Organization.

Fletcher, L.B. (1992) 'Rethinking world food, trade, aid and food security issues for the 1990s: An introductory essay', in L.B. Fletcher (ed.) *World Food in the 1990s: Production, Trade and Aid*, Boulder, Colo.: Westview Press.

Flew, A. (ed.) (1970) Malthus, T.R., *An Essay on the Principle of Population and a Summary View of the Principle of Population*, Harmondsworth: Penguin.

Foote, K.A., Hill, K.H. and Martin, L.G. (eds) (1993) *Demographic Change in Sub-Saharan Africa*, Washington, DC: National Academy Press.

Freedom House (1995) *Freedom in the World: The Annual Survey of Political Rights and Civil Liberties, 1994–1995*, New York: Freedom House.

Gastil, R. (ed.) (1989) *Freedom in the World 1988–89*, New York: Freedom House.

Geertz, C. (1963) *Agricultural Involution: The Process of Ecological Change in Indonesia*, Berkeley and Los Angeles: University of California Press.

Gilland, B. (1993) 'Cereals, nitrogen and population: An assessment of the global trends', *Endeavour, New Series* 17, 2: 84–8.

Glantz, M.H. (1991) 'On the interactions between climate and society', in K. Davis and M.S. Bernstam (eds) *Resources, Environment, and Population*, Supplement to *Population and Development Review* 16, New York: Oxford University Press.

Gleave, B. (1994) 'Population density, population change, agriculture and the environment in Tropical Africa', in B. Zaba and J. Clarke (eds) *Environment and Population Change*, Liège: Derouaux Ordina Editions.

Goswami, O. (1990) 'The Bengal famine of 1943: Re-examining the data', *Indian Economic and Social History Review* 27, 4: 445–63.

Greeley, M. (1992) 'Agricultural biotechnology, poverty and employment', paper prepared for the Technology and Employment Branch of the ILO in association with ODA/IDS Research Programme IV, Brighton: Institute of Development Studies.

Greenspan, A. (1994) 'Apocalypse when? Population growth and food supply in South Asia', *Asia-Pacific Population and Policy* 31, Honolulu: East-West Center.

Griffin, M. (1993) 'Controlling the cuckoo', *Ceres* 26, 3: 32–5.

Grigg, D. (1993) *The World Food Problem*, 2nd edn, Oxford: Basil Blackwell.

Hadley Centre (1993) 'Progress Report 1990–1992 and Future Programme of Research', Hadley Centre for Climate Prediction and Research, Bracknell, United Kingdom.

Hammarskjöld, M., Egerö, B. and Lindberg, S. (eds) (1992) *Population and the Development Crisis in the South*, Lund: PROP Publication Series.

Harrison, P. (1992) *The Third Revolution: Environment, Population and a Sustainable World*, London: I.B. Tauris.

Harriss, B., Guhan, S. and Cassen, R.H. (eds) (1992) *Poverty in India*, Bombay: Oxford University Press.

Hayami, Y. and Ruttan, V.R. (1987) 'Population growth and agricultural poverty', in D.G. Johnson and R.D. Lee (eds) *Population Growth and Economic Development: Issues and Evidence*, Madison: University of Wisconsin Press.

Heilig, G.K. (1994) 'Neglected dimensions of global land-use change: Reflections and data', *Population and Development Review* 20, 4: 831–59.

Henderson-Sellers, A. (1994) 'Numerical modelling of global climates', in N. Roberts (ed.) *The Changing Global Environment*, Oxford: Basil Blackwell.

Hendry, P. (1991) 'Food and population beyond five billion', *Population Bulletin* 43, 2, Washington, DC: Population Reference Bureau.

Higgins, G.M., Kassam, A.H., Naiken, L. and Shah, M.M. (1981) 'Africa's agricultural potential', *Ceres* 14, 5: 13–21.

Homer-Dixon, T.F., Boutwell, J.H. and Rathjens, G.W. (1993) 'Environmental change and violent conflict', *Scientific American* 268, 2: 16–23.

Hung, N.M., Jamieson, N.L. and Rambo, A.T. (1991) *Environment, Natural Resources, and the Future Development of Laos and Vietnam: Papers from a Seminar*, Fairfax, VA: Indochina Institute at George Mason University.

Huxley, A. (1949) 'The double crisis', *Science News Letter*, 26 March: 1–4, Paris: United Nations Educational, Scientific and Cultural Organization.

ICRAF (1993) *ICRAF: A Global Agenda*, Nairobi: International Centre for Research in Agroforestry.

IPCC (1990a) *Climate Change: The IPCC Scientific Assessment*, Cambridge: Cambridge University Press.

——— (1990b) *Scientific Assessment of Climate Change*, Geneva: World

Meteorological Organization/United Nations Environment Programme.

——— (1992) *1992 IPCC Supplement: Scientific Assessment*, Geneva: World Meteorological Organization/United Nations Environment Programme.

——— (1994) *Radiative Forcing of Climate Change*, Geneva: World Meteorological Organization/United Nations Environment Programme.

IRRI (1993) *IRRI Rice Almanac*, Manila: International Rice Research Institute.

Jaeger, W.K. (1992) *The Effects of Economic Policies on African Agriculture*, Washington, DC: The World Bank.

Jefferson, T. (1804/1993) 'Letter to Jean Baptiste Say (February 1, 1804)', reprinted in *Population and Development Review* 19, 1: 180–1.

Kane, S., Reilly, J. and Tobey, J. (1992) 'An empirical study of the economic effects of climate change on world agriculture', *Climatic Change* 21, 1: 17–35.

Kates, R. (1993) 'Divisions in the hunger industry', *The Times Higher*, London, April.

Kellman, M. (1987) *World Hunger, A Neo-Malthusian Perspective*, New York: Praeger.

Keyfitz, N. (1991) 'Population and development within the ecosphere: One view of the literature', *Population Index* 57, 1: 5–22.

Keynes, J.M. (1919) *The Economic Consequences of the Peace*, London: Macmillan.

King, M. (1990) 'Health is a sustainable state', *The Lancet* 336: 664–7.

——— (1991) 'Human entrapment in India', *National Medical Journal of India* 4, 4: 196–201.

——— (1992) 'Escaping the demographic trap', in M. Hammarskjöld, B. Egerö and S. Lindberg (eds) *Population and the Development Crisis in the South*, Lund: PROP Publication Series.

Kumar, B.G. (1990) 'Ethiopian famines 1973–1985: A case study', in J. Drèze and A. Sen (eds), *The Political Economy of Hunger* 2, Oxford: Clarendon Press.

Kumar, B.G. and Stewart, F. (1992) 'Tackling malnutrition: What can targeted nutritional interventions achieve?', in B. Harriss, S. Guhan and R.H. Cassen (eds) *Poverty in India*, Bombay: Oxford University Press.

Kutzner, P.L. (1991) *World Hunger, A Reference Handbook*, Santa Barbara and Oxford: ABC-CLIO.

Larson, W., Pierce, F. and Dowdy, R. (1983) 'The threat of soil erosion to long-term crop production', *Science* 219: 458–65.

Lee, R.D. (1991) 'Long-run global population forecasts: A critical appraisal', in K. Davis and M.S. Bernstam (eds) *Resources, Environment and Population*, Supplement to *Population and Development Review* 16, New York: Oxford University Press.

Leibenstein, H. (1957) *Economic Backwardness and Economic Growth*, New York: John Wiley.

Lele, U. and Stone, S.S. (1989) 'Population pressure, the environment and agricultural intensification', Managing Agricultural Development in Africa, Discussion Paper 4, Washington, DC: The World Bank.

Lipton, M. (1993) 'Don't neglect reform of agriculture', *Weekly Mail and Guardian*, Johannesburg, 6–12 August.

Malthus, T.R. (1798) *An Essay on the Principle of Population as it Affects the Future Improvement of Society with Remarks on the Speculations of Mr. Godwin, M. Condorcet, and Other Writers*, London: Printed for J. Johnson, in St Paul's Church-yard.

——— (1830/1970) *A Summary View of the Principle of Population*, London: John Murray, Albemarle-Street; reprinted in A. Flew (ed.) (1970) Malthus, T.R., *An Essay on the Principle of Population and a Summary View of the Principle of Population*, Harmondsworth: Penguin.

Matthews, A. (1993) 'Necessity neglected', *Ceres* 26, 3: 24–8.

McNicoll, G. (1992) 'The United Nations' long-range population projections', *Population and Development Review* 18, 2: 333–40.

Mendelsohn, R., Nordhaus, W.D. and Shaw, D. (1994) 'The impact of global warming on agriculture: A Ricardian analysis', *American Economic Review* 84, 4: 753–70.

Menken, J. and Campbell, C. (1992) 'Age-patterns of famine-related mortality increase: Implications for long-term population growth', *Health Transition Review* 2, 1: 91–101.

Merrick, T.W. (1991) 'Population pressures in Latin America', *Population Bulletin* 41, 3, Washington, DC: Population Reference Bureau.

Messer, E. and Heywood, P. (1990) 'Trying technology, neither sure nor soon', *Food Policy* 15, 4: 336–45.

Mitchell, D.O. and Ingco, M.D. (1993) *The World Food Outlook*, Washington, DC: The World Bank.

Mulder, D.W., Nunn, A.J., Wagner, H., Kamali, A. and Kengeya-Kayondo, J.F. (1994) 'HIV-1 incidence and HIV-1-associated mortality in a rural Ugandan population cohort', *AIDS* 8, 1: 87–92.

Museveni, Y.K. (1992) *What is Africa's Problem?*, Kampala: NRM Publications.

Myers, N. (1991) *Population, Resources and the Environment: The Critical Challenges*, New York: United Nations Population Fund.

Nahaylo, B. (1994) 'Population displacement in the former Soviet Union', *Refugees* 4, 98: 3–8.

Nelson, R.R. (1956) 'A theory of the low-level equilibrium trap in undeveloped economies', *American Economic Review* 46, 5: 894–908.

NFU (1994) 'Real choices', Discussion Document, Long Term Strategy Group, London: National Farmers' Union.

Nickum, J.E. (1990) 'Volatile waters: Is China's irrigation in decline?', in T.C. Tso (ed.) *Agricultural Reform and Development in China*, Beltsville, Md: Ideals, Inc.

Ottichilo, W.K. (1993) 'Operational application of satellite remote sensing in an early warning system in east African countries', *International Training Centre Journal* 3: 261–6.

Oucho, J.O. and Gould, W.T.S. (1993) 'Internal migration, urbanization, and population distribution', in K.A. Foote, K.H. Hill and L.G. Martin (eds) *Demographic Change in Sub-Saharan Africa*, Washington, DC: National Academy Press.

Parry, M. (1990) *Climate Change and World Agriculture*, London: Earthscan Publications Ltd.

Paulino, L.A. and Tseng, S.S. (1980) 'A comparative study of FAO and USDA data on production, area and trade of major food staples', Research Report 19, Washington, DC: International Food Policy Research Institute.

Pearce, F. (1992) 'Mirage of the shifting sands', *New Scientist* 136, 1851: 38–42.

Peng, X. (1987) 'Demographic consequences of the Great Leap Forward in China's provinces', *Population and Development Review* 13, 4: 639–70.

Pingali, P. and Binswanger, H.P. (1991) 'Population density and farming systems', in R.D. Lee, W.B. Arthur, A.C. Kelley, G. Rodgers and T.N. Srinivasan (eds) *Population, Food, and Rural Development*, Oxford: Clarendon Press.

Pinstrup-Andersen, P. (1994a) 'World food trends and future food security', Food Policy Report, Washington, DC: International Food Policy Research Institute.

——— (1994b) 'World food trends and future food security: Meeting tomorrow's food needs without exploiting the environment', Food Policy Statement 18, Washington, DC: International Food Policy Research Institute.

Plucknett, D.L. (1993) 'Science and agricultural transformation', IFPRI Lecture Series 1, Washington, DC: International Food Policy Research Institute.

Popkin, B.M. (1993) 'Nutritional patterns and transitions', *Population and Development Review* 19, 1: 138–57.

Population Action International (1992) *Population and the Environment: Impacts in the Developing World*, Washington, DC: Population Action International.

Postel, S. (1989) 'Water for agriculture: Facing the limits', *Worldwatch Paper 93*, Washington, DC: Worldwatch Institute.

——— (1992) *Last Oasis, Facing Water Scarcity*, New York: W.W. Norton.

——— (1994) 'Irrigation expansion slowing', in L.R. Brown, H. Kane and D.M. Roodman (eds) *Vital Signs 1994, 1995*, London: Earthscan Publications Ltd.

Preston, S.H. (1975) 'The changing relation between mortality and level of economic development', *Population Studies* 29, 2: 231–48.

Ravi Kanbur, S.M. (1991) 'Global food balances and individual hunger: Three themes in an entitlements-based approach', in J. Drèze and A. Sen (eds) *The Political Economy of Hunger* 1, Oxford: Clarendon Press.

Robinson, J. (1941) 'Rising supply price', *Economica* 8, 29: 1–8.

Roodman, D.M. (1994a) 'Global temperature rises slightly', in L.R. Brown, H. Kane and D.M. Roodman (eds) *Vital Signs 1994, 1995*, London: Earthscan Publications Ltd.

——— (1994b) 'Carbon emissions unchanged', in L.R. Brown, H. Kane and D.M. Roodman (eds) *Vital Signs 1994, 1995*, London: Earthscan Publications Ltd.

Rosengrant, M.R. (1993) 'Markets in tradable water rights could improve irrigation productivity in developing countries', *IFPRI Report* 15, 2, Washington, DC: International Food Policy Research Institute.

Rosenzweig, C. and Parry, M. (1994) 'Potential impact of climate change on world food supply', *Nature* 367, 6459: 133–8.

Russell, E.J. (1949) 'The way out', *Science News Letter*, 2 April 1949: 5–8, Paris: United Nations Educational, Scientific and Cultural Organization.

Russell, S.S. (1993) 'International migration', in K.A. Foote, K.H. Hill and L.G. Martin (eds) *Demographic Change in Sub-Saharan Africa*, Washington, DC: National Academy Press.

Ryan, M. (1994) 'CFC production continues to drop', in L.R. Brown, H. Kane and D.M. Roodman (eds) *Vital Signs 1994, 1995*, London: Earthscan Publications Ltd.

Saether, A. (1993) 'Otto Diederich Lütken: 40 years before Malthus?', *Population Studies* 47, 3: 511–13.

Salter, R.M. (1948) 'World soil and fertilizer resources in relation to food needs', *Chronica Botanica* 11, 4: 226–35.

Sarma, J.S. and Gandhi, V.P. (1990) 'Production and consumption of foodgrains in India: Implications of accelerated economic growth and poverty alleviation', Research Report 81, Washington, DC: International Food Policy Research Institute.

Scott, L. and Vollrath, T. (1992) 'Global competitive advantages and overall bilateral complementarity in agriculture: A statistical review', Economic Research Service, Statistical Bulletin 850, Washington, DC: United States Department of Agriculture.

Seaman, J. (1993) 'Famine mortality in Africa', in J. Swift (ed.) 'New approaches to famine', *IDS Bulletin* 24, 4: 27–31.

Sen, A. (1981) *Poverty and Famines: An Essay on Entitlement and Deprivation*, Oxford: Clarendon Press.

——— (1990) 'Food, economics and entitlements', in J. Drèze and A. Sen (eds) *The Political Economy of Hunger* 1, Oxford: Clarendon Press.

Shirahatti, P.P. (1989) 'Agriculture: long term prospects', in N. Bhaskara Rao (ed.) *India 2021*, Baroda: Operations Research Group.

Simon, J.L. (1981) *The Ultimate Resource*, Oxford: Martin Robertson.

——— (1992) *Population and Development in Poor Countries, Selected Essays*, Princeton, NJ: Princeton University Press.

Skeldon, R. (1984) *Migration in South Asia: An Overview*, Bangkok: Economic and Social Commission for Asia and the Pacific.

Smil, V. (1991) 'Population growth and nitrogen: An exploration of a critical existential link', *Population and Development Review* 17, 4: 569–601.

——— (1994) 'How many people can the earth feed?', *Population and Development Review* 20, 2: 255–92.

Smith, A. (1776) *An Inquiry into the Nature and Causes of the Wealth of Nations*, vol. 2, London: Printed for W. Strahan and T. Cadell in the Strand.

Starke, L. (1993) 'Population growth sets another record', in L.R. Brown, H. Kane and E. Ayres (eds) *Vital Signs 1993*, London: Earthscan Publications Ltd.

State Statistical Bureau of the People's Republic of China (1994) *Statistical Yearbook of China 1994*, Beijing: China Statistical Publishing House.

Sugden, D. and Hulton, N. (1994) 'Ice volumes and climate change', in N. Roberts (ed.) *The Changing Global Environment*, Oxford: Basil Blackwell.

Sukhatme, P.V. and Margen, S. (1982) 'Autoregulatory homeostatic nature of energy balance', *American Journal of Clinical Nutrition* 35: 355–65.

Teklu, T., Braun, J. von and Zaki, E. (1991) 'Drought and famine relationships in Sudan: Policy implications', Research Report 88, Washington, DC: International Food Policy Research Institute.

Tomkins, A. and Watson, F. (1989) 'Malnutrition and infection: A review', United Nations ACC/SCN State-of-the-Art Series, Nutrition Policy Discussion Paper 5, London: London School of Hygiene and Tropical Medicine.

Tyagi, D.S. (1990) *Managing India's Food Economy*, New Delhi: Sage.

United Nations (1989) *World Population Prospects 1988*, New York: United Nations.

——— (1991) *World Urbanization Prospects 1990*, New York: United Nations.

——— (1993) *World Population Prospects: The 1992 Revision*, New York: United Nations.

——— (1994) *World Population Prospects: The 1994 Revision – Annex Tables*, New York: United Nations.

United Nations Environment Programme (1992) *World Atlas of Desertification*, London: Edward Arnold.

USDA (1992) 'Agricultural Resources: Cropland, Water and Conservation', Economic Research Service, AR-27, Washington, DC: United States Department of Agriculture.

——— (1993a) 'Former USSR international agriculture and trade report', Economic Research Service, RS-93-1, Washington, DC: United States Department of Agriculture.

——— (1993b) 'China international agriculture and trade report', Economic Research Service, RS-93-4, Washington, DC: United States Department of Agriculture.

——— (1993c) 'Industrial uses of agricultural materials', Economic Research Service, IUS-1, Washington, DC: United States Department of Agriculture.

Virmani, S.S., Khush, G.S. and Pingali, P.L. (1994) 'Hybrid rice for tropics: Potentials, research priorities and policy issues', in R.S. Paroda and Mangala Rai (eds) *Hybrid Research and Development of Major Cereals in the Asia-Pacific Region*, Bangkok: Food and Agricultural Organization.

Watkins, S.C. and Menken, J. (1985) 'Famines in historical perspective', *Population and Development Review* 11, 4: 647–75.

Webb, P., Braun, J. von and Yohannes, Y. (1992) 'Famine in Ethiopia: Policy implications of coping failure at national and household levels', Research Report 92, Washington, DC: International Food Policy Research Institute.

Woodward, F.N. (1950) 'Creatable resources: The development of new resources by applied technology', *Proceedings of the United Nations Scientific*

Conference on the Conservation and Utilization of Resources 1: 131–5, New York: United Nations.

World Bank (1986) *Poverty and Hunger: Issues and Options for Food Security in Developing Countries*, Washington, DC: The World Bank.

——— (1990) 'Agricultural biotechnology: The next "Green Revolution"?', World Bank Technical Paper 133, Washington, DC: The World Bank.

——— (1992) *Development and the Environment*, World Development Report 1992, New York: Oxford University Press.

——— (1993a) *Global Economic Prospects and the Developing Countries 1993*, Washington, DC: The World Bank.

——— (1993b) *Sub-Saharan Africa: From Crisis to Sustainable Growth*, Washington, DC: The World Bank.

——— (1993c) 'A strategy for managing water in the Middle East and North Africa', paper presented by the Water Resource Management Unit, Washington, DC: The World Bank.

——— (1993d) *Arab Republic of Egypt: An Agricultural Strategy for the 1990s*, Washington, DC: The World Bank.

World Bank and United Nations Development Programme (1989) *Africa's Adjustment and Growth in the 1980s*, Washington, DC: The World Bank.

World Resources Institute (1990) *World Resources 1990–91*, New York: Oxford University Press.

——— (1992) *World Resources 1992–93*, New York: Oxford University Press.

——— (1994) *World Resources 1994–95*, New York: Oxford University Press.

Wrigley, E.A. and Schofield, R.S. (1981) *The Population History of England 1541–1871: A Reconstruction*, London: Edward Arnold.

Wyles, J. (1993) 'Compensation: The consolation prize', *Ceres* 26, 3: 28–31.

INDEX

access to food 2
accuracy of data 24–5, 27
Afghanistan 50, 66, 84, 147, 180–1
ageing population 104, 195
aggregation of data 26, 56, 103
agricultural data 25–7
AIDS 103, 175
Alexandratos, N. 116
Algeria 44, 55, 86, 89, 178
An Essay on the Principle of Population (Malthus) 3–5, 23, 186
Anderson, J.R. 146
Angola 55, 85
anti-Malthusianism 6–11, 22
Argentina 44, 55, 66, 84–5, 90, 99, 121, 129, 167, 192
Armenia 193
Australia 11, 29, 84, 144, 159, 167, 196–9
Austria 194
availability: calorie 34, 36–9, 93, 95; cereal 40, 46–7; data construction and 2, 203; developing countries 56–7, 206; famine and 74, 104; Far East 185; food 34, 201; Latin America 191–2; Middle East 178; security and 52–3, 55; South Asia 183
Azerbaijan 193

balance between population and food 1–2, 4, 11, 18–22
Bangladesh 15, 39, 44–6, 56, 73, 94, 128, 139, 180–4
barley 26, 139, 179, 187
basal metabolic rate (BMR) 35
Binswanger, H.P. 9
biotechnology 97, 158–60, 166, 197
birth rate: death rate/ 5, 12–13, 31; decline 20–1, 50; developing countries 48; education 162; estimation of 25; Far East 32, 186; food crises and 69, 73; Latin America 32, 191; Middle East 32, 179; regional 31–3; South Asia 32, 182–3
Bolivia 39, 191–2
Boserup, Ester 7–8, 10
Botswana 31, 173
Brazil 44, 55, 85, 160, 191–2
Brown, Lester R. 11, 16, 72, 187–8
Burundi 85

calorie: intake 27, 33–9, 41–2; per capita supply 47–51
Cambodia 185–6
Cameroon 174
Canada 29, 45, 69–70, 82, 84, 191, 196–9
carbon dioxide, atmospheric 137, 139, 141–2
Centro Internacional de Mejoramiento de Maiz y Trigo (CIMMYT) 64, 158
Chechnya 193
Chile 55, 192
China 29, 33, 39, 41, 55, 73, 81, 85, 89, 94, 105, 113, 117–19, 139,

150, 153, 156, 158, 160, 166–7, 184–90, 200, 205; 1959–64 food disaster 68–71
chlorofluorocarbon gases (CFCs) 137–8, 143
climatic change 14–15, 17, 139–40
coarse grains 26
Colombia 192
Common Agricultural Policy (CAP) 84, 195–6
communications 20–1, 95, 161–3
conclusions, forecasts, caveats, tempered hope 201–9; caveats 205–8; conclusions 201–3; forecasts 203–5; tempered hope 208–9
Congo 174
conservation 149, 154, 179, 186, 206
Conservation Reserve Program 198
Consultative Group on International Agricultural Research (CGIAR) 159–60, 165, 172
consumption: direct/indirect 27, 46–7; Europe/former Soviet Union 109–10, 192–4; Far East 47, 107–9, 111–14, 206; global 1–2, 100, 113; Latin America 107, 109, 111–13, 206; Middle East 107–8, 110–12, 113–14, 178, 206; North America/ Oceania 47, 107, 109–11; per capita 39–47, 133–4, 202–6; and production differential 46; regional 40; South Asia 107–8, 111–13, 180–1, 183–4, 206; sub-Saharan Africa 47, 108–11, 170, 173, 177, 206; surveys 34; warfare and 50
contraception 20
credit facilities 174
crises and famines 17, 68–75, 96, 170
Crookes, Sir William 5
crop improvement 20–1, 64–5, 158–61, 168, 179, 190, 197, 204
cropland: decrease 15, 62–3, 75, 81–4, 94, 99, 187–8; effect of sea level rise 15, 139; Europe/ former Soviet Union 194; expansion 11, 59–60, 122, 131–2, 170, 204; fixed area 5, 13, 125, 183; harvested area 125–9, 131–4; idle 69, 73, 88–9, 131, 198; increase 70; North America/Oceania 196, 198; potential 115–16, 120–1, 147–9, 167–8, 174; relative to production 59–63; supply 19–21; supply and food security 51
Crosson, P. 146
Cuba 44, 55, 146
Czechoslovakia 194

dams 153
data of sample countries 210–11
death rate: balance of food and population 4, 20, 104; decline 48; education and 162; estimation of 25; excess 17–18, 68, 71, 73–4; hunger and 56–7; increase 14, 18; population projections and 104; regional 32–3; size of population/ 3–4
deforestation 14–15, 117, 144, 147–8, 185, 191
demand for food 6, 13–14, 19, 89, 94, 100–14, 167–8; supply/ 125–35, 203–5
demand and supply in future 100–35; cereal demand 101–14; cereal supply 115–24; overview of scenario 132–3; weighing of 125–32
demographic data 24–5, 31–2
Denmark 194
denudation of land 144
desertification 13–14, 144
developing countries: and food aid 89–91; biotechnology 158–60; birth rate 13, 32, 48; conservation 149; data collection 25–6; education 162; famine 17; fertilizers 119, 156; food security 55–6; food storage 161; global warming and 140–2; imports 168; pesticide production 157;

population growth 13, 15, 104; prices 98, 164; production 84–6, 92–3; in sample 31; water 118; yields 99
diet 33–4, 37–9, 41, 47, 94–5, 110, 133, 181–2, 202
disease 4, 37, 174–5
distribution 2, 11, 35, 164–6, 205
diversity of food 64, 91–5
drought 68–9, 73, 87–9, 97, 140, 142, 150, 155, 159, 166–7, 171, 173, 181, 197, 207

economic growth 89, 101, 105–6, 109–10, 168, 173
education 20–1, 162–3
Egypt 15, 44–6, 56, 119, 122, 139, 158, 178–80
Ehrlich, Paul R. 11
El Niño-Southern Oscillation (ENSO) 69, 73, 167, 207
El Salvador 192
electricity 161, 167
emergency relief 51–2, 165
employment patterns 95
Engels, Friedrich 7
environmental change 14–16, 136, 143–9
environmental data 24, 27–8
estimation, statistical 24–5, 87
Ethiopia 39, 46, 71, 74, 85, 174
Europe 15, 46, 48, 140, 149, 157, 162, 164–5, 187, 205
Europe/former Soviet Union: calorie availability 36–9, 43; cereal consumption 47; cereal production, consumption and trade 40–1, 44; comparison with North America/Oceania 197–8; consumption 107, 109–11; demand/supply 126, 130–4; demographic and socio-economic data 32–3; exports 168; fertilizers 119; food security 53–5; future prospects 192–6; harvest variability 87–8, 207; population projection 102–3; cropland 115–16, 120; production 40, 81–3, 92–4, 191–6; as region 29–30; trade 89–90, 161; yields 98, 121–4
European Union (EU) 44–5, 83–4, 90–1, 97, 195–8, 204
exports 41, 44, 81, 84–5, 90, 97, 130–4, 154, 167, 178, 191–3, 196–7, 204–5

fallow periods 8, 144, 147, 170, 185
family planning programmes 50, 179
famine 4, 14, 17, 18, 19, 68–75, 96, 104, 171, 177–8, 181, 195, 207
Far East: and food aid 44; birth rate 32–3, 50; calorie availability 36, 38–9, 43; cereal production, consumption and trade 40–1; consumption 47, 107–9, 111–14, 206; demand 167–8; demand/supply 126, 128–30, 133–4; demographic and socio-economic data 32–3; economic growth 89, 105–6; fertilizers 118; food security 54–5; future prospects 184–90; harvest variability 87–8; imports 205; irrigation 117–18, 150, 153, 204; population projection 102–3; potential cropland 115–16; production 40, 77, 79–83, 92–5; as region 29–30; trade 91; yields 121–4
fertility rates 32–3, 103–4
fertilizers 7, 16, 21, 64, 116, 118, 127–8, 130, 132, 134, 138, 148–9, 155–7, 161–2, 167–8, 171–2, 174, 179, 182–4, 186, 190, 192, 204
flooding 14–15, 64, 68, 73, 140
food aid 40, 44–6, 51, 56, 71–2, 84, 89–91, 95, 97–8, 127, 165, 170, 172–3, 177–8, 180, 183, 185, 191, 206
Food and Agriculture Organization (FAO) 25–6, 34–8, 92, 97, 115, 191
food as data category 25, 92
food security index (FSI) 52–5, 185, 210–11

former Soviet Union 28–9, 44, 88, 192–4, 205, *see also* Europe/former Soviet Union
France 128, 159, 195
fruit 25, 92, 95, 199
future potentials and constraints 136–69; farm inputs, crop developments and research 155–61; food prices 166–8; global atmospheric changes 136–43; institutions and markets 163–6; physical and human capital 161–3; soil and water 143–55
future regional vignettes 170–200; Europe/former Soviet Union 192–6; Far East 184–90; Latin America 191–2; Middle East 177–80; North America/Oceania 196–9; South Asia 180–4; sub-Saharan Africa 170–7

Gabon 174
GDP forecasts 102, 104
Geertz, Clifford 10
General Agreement on Tariffs and Trade (GATT) 163–4, 167, 195, 198, 205
Georgia 193
Germany 119
Ghana 176
Global Assessment of Soil Degradation (GLASOD) 144–7
global population 1, 12–13, 101–5
global warming 14–15, 28, 88, 136–42, 168, 208
GNP per capita 32
gravity flooding 150
green revolution 7, 16, 64–9, 97, 155, 160, 172, 181, 203
greenhouse gases (GHGs) 15–16, 137–9, 142
growth, population: and consumption 106–14; data 25; declining rate of 20–1; and demand 107, 119, 125–8, 132–4; estimates 31–3, 101–5 Europe/former Soviet Union 193; food crises and 69, 73–5; food supply and 47–51; global 1, 5, 10, 101–5; market efficiency and 6–7; Middle East 178–80; per capita calorie supply and 48–51; production and 58–9, 81, 83, 191–2, 201–3, 205, 207–9; rates of 32; 31; South Asia 182–4; sub-Saharan Africa 170–2, 177
Guatemala 192

Haiti 55, 146, 192
harvest: failure 74; global 17, 58–60, 62–3, 70; irrigation and 117; stability 64; variability 87–9, 96, 142, 179, 206–7
high yielding varieties (HYVs) 7, 21, 64–6, 68, 97, 172, 181
Honduras 192
human capital 161–3, 168
humanitarianism 44, 46, 56, 180, 206
Hungary 194
hunger 33–9, 56, 201–2, 206

immunization 57, 162
imports 44, 52, 55, 70, 84–5, 89–90, 125–34, 163–4, 167, 170, 172–3, 177–81, 183–5, 187, 189–92, 202, 205–7
income 33, 47, 52–3, 85, 89, 94, 96, 101–2, 105, 133, 206
India 29, 44, 71, 85, 90, 94, 117–18, 150–1, 153, 156, 158, 160, 166, 181–4
Indochina 115
Indonesia 10, 84, 115, 139, 157, 160, 183, 185–7, 189
innovation 7–10, 20–1, 162
institutions and markets 163–6
integrated pest management (IPM) 157–8
intensive/extensive cropping 8, 10
Intergovernmental Panel on Climate Change (IPCC) 136, 138–9, 142
International Crops Research Institute for the Semi-Arid Tropics (ICRISAT) 159

International Monetary Fund (IMF) 173
International Rice Research Institute (IRRI) 64, 158
Iran 44, 50, 117, 128, 180–1, 184
Iraq 55, 122, 147, 178–80
irrigation 13–15, 21, 64–5, 87, 116–18, 122, 144, 149–54, 172, 179, 182, 184, 186, 188, 190, 204
Italy 195

Japan 29, 33, 41, 44, 48, 65, 89, 128, 164, 185–6, 189–90

Kazakhstan 140, 193–4
Kenya 31, 39, 67, 176
Keynes, John Maynard 5

land: degradation 13–14, 28, 143–8, 168, 185, 191; potential 115–16, 120–2, 143–4, 147–9, quality 144; use 8–10, 13, 190
Latin America: and food aid 44–5; birth rate 32, 50; calorie availability 36, 37–9, 43; cereal availability 40, 46; cereal production, consumption and trade 40–1, 44; consumption 107, 109, 111–13, 206; demand/supply 126, 128–9, 134; demographic and socio-economic data 32–3; economic growth 105; fertilizers 119, 192; food security 54–5; future prospects 191–2; harvest variability 87–8; imports 44, 205; population projection 102–3; potential cropland 115–16; production 40, 76, 78, 81–3, 92–5; as region 29–30; soil erosion 147; trade 90–91; yields 121–4
life expectancy 31, 32–3
linseed 94
livestock feed 27, 40, 46, 56, 133, 156, 189, 193
Lütken, Friderich C. 6

machinery, agricultural 161, 167, 171
Madagascar 118, 147
maize 26, 41, 64, 139, 160, 172–3, 197–9
Malawi 50
Malaysia 86, 185, 187
Malthus, Thomas Robert 3–5, 7, 11, 61, 186, 209
markets 6–7, 163–6, 174, 198, 200
meat 92, 95, 110, 132
methane 137
Mexico 29, 44, 55, 86, 89–90, 117, 146, 151, 159–60, 192, 199
Middle East: birth rate 50; calorie availability 36–8, 42; cereal availability 46, 202; cereal production, consumption and trade 40–4; consumption 107–8, 110–11, 113–14, 206; demand/supply 125–7, 133–4; demographic and socio-economic data 32–3; economic growth 89, 105; fertilizers 119, 157; and food aid 45–6; food security 54–5, 178; future prospects 177–80; harvest variability 87–8; imports 41–4, 70, 205; irrigation 118, 122, 150, 154, 178–9; per capita food supply 49; population projection 102–3; potential cropland 116; production 40, 77, 79–83, 93; as region 29–31; soil erosion 148; trade 90–1; yields 121–4
migration: China 187; in- 33, 50, 185; long-distance 117, 174; out- 174, 191; population density and 9–10, 20, 174; rural to rural 174; rural to urban 105, 176, 195
millet 26, 41, 139, 159–60, 172, 181
Moldova 193
Morocco 29, 122, 178, 180
Mozambique 46, 50, 55, 85, 147, 176
multiple cropping 59, 64, 120–2, 132, 182, 186, 188, 204
Myanmar 55, 85, 185–7, 190
Myers, Norman 11, 16

neo-Malthusianism 3, 6, 11–20, 58, 83, 87, 95–6, 104, 201, 203, 207–8
Netherlands 15, 119
Niger 54, 147
Nigeria 118, 173, 177
nitrogen 118–19, 155–6, 182, 204
nitrous oxide 16, 138
non-governmental organization (NGO) 166
North America 11, 15, 46, 48, 149, 157, 162, 164, 187, 202, 204
North America/Oceania: calorie availability 36–9, 43; cereal availability 46; cereal production, consumption and trade 40–1; consumption 47, 107, 109–11; demand/supply 126; demographic and socio-economic data 32–3; exports 41, 97, 130–1, 134, 167–8; and food aid 45; food security 53–5; future prospects 196–9; harvest variability 88–9, 142, 206–7; population projection 102–3; production 40, 77, 79–83, 92–3; as region 29–30; trade 41, 56, 90–1, 161; water 150, 154–5; yields 99, 121–4
North American Free Trade Agreement (NAFTA) 199
North Korea 114, 119, 128, 186
nuclear accident 166

oats 26
Oceania *see* North America/Oceania
overgrazing of animals 144
ozone depletion 14–15, 136–8, 142–3

Pakistan 50, 65–6, 117, 151, 181, 183–4
Peru 39, 55, 191
pessimists and optimists 1–23; anti-Malthusianism 6–11; history and theory 3–6; modern neo-Malthusianism 11–17; population and food 1–3; question of balance 18–22
pest movement detection 162
pesticides 21, 64, 157–8, 161, 167
Philippines 185, 189
phosphate 156–7, 182
physical capital 161–3, 168
Pingali, P. 9
plastics, synthetic 161
Poland 159, 194, 196, 198
pollution 15, 138, 144
Postel, S. 152
potash 156, 182
potatoes 34, 94, 139, 160
poverty 11, 47, 56, 171, 181
prawns 93
prices 6, 14, 17, 70–2, 74–5, 83–5, 90, 97–9, 129–30, 141, 149, 163–9, 172–3, 184, 191–2, 194–6, 199, 202, 205, 207
production: AIDS and 175; availability and diversity 91–5; capacity 3, 5; constraints 13; consumption and 56, 165; consumption, trade and aid 39–47; Europe/former Soviet Union 81–3, 92–4, 194–5; Far East 77, 79–83, 92–5, 184, 187–8; fertilizer 119, 155; food diversity and 90–4; global 1–2, 40, 58–63, 65–8; institutions and 163; land degradation and 147; Latin America 76, 78, 81–3, 92–5; Middle East 75, 77, 79–83, 93; North America/Oceania 77, 79–83, 92–3; per capita 58–63, 68–70, 73–4, 87, 96–7, 202; pesticides 157; population growth and 11, 16–17, 21–2, 95–7, 168; regional cereal 75–87; South Asia 76, 78, 181; sub-Saharan Africa 76, 78, 170–1, 173, 176; water and 153–4
projections, population 12–13, 101–5, 114, 177, 179–80, 182, 184, 186, 191–2
protectionism 163
protein intake 33–4, 38–9

quantitative data 24; problems of 25–8

radiation 14–15, 143
rainfall 51, 75, 119, 140, 144, 154, 162, 171, 182, 194, 207
rapeseed 93, 198
regions: dividing world into 28; population distribution 29–31
requirements, calorie 35–8, 101
research 20, 64, 68, 158–60, 165–6, 172, 174, 179
resources, agricultural 14, 115–19
rice 15–16, 26, 41, 93, 139, 143, 148, 157–8, 160, 185–7, 189; HYVs 7, 64–5, 172, 181
Roodman, D.M. 137
root and tuber crops 41, 93, 96, 172
Russia 193, 198
Rwanda 85, 175
rye 26, 158

Sahel 71, 118, 140, 144, 147–8, 174
salinization 14–15, 145, 150–1, 179
satellite remote-sensing 162, 204
Saudi Arabia 44, 67, 85, 122, 179–80
science 7, 11, 20
scientific research 7, 11, 22
sea level rises 15, 139–40
security, food 51–6, 99, 154, 165–6, 170, 175, 178, 181, 185, 191, 193, 196, 201–2, 206–7
Senegal 176
Simon, Julian L. 6
Smil, V. 157
Smith, Adam 6, 74
socio-economic data 31–3
soil erosion 10, 13–14, 28, 140, 144–5, 148, 170
Somalia 55, 85
sorghum 26, 41, 139, 172–3, 181
South Africa 39, 55, 118, 147, 157, 175–6
South Asia: birth rate 32, 50; calorie availability 36, 38–9, 43, 48–9; cereal production, consumption and trade 40, 41, 44; consumption 107–8, 111–13, 206; demand/supply 126, 128, 134; demographic and socio-economic data 32; drought 69–70, 181; economic growth 105, 108; fertilizers 119, 184; and food aid 44–6; food security 54–5; future prospects 180–4; harvest variability 87; imports 205; irrigation 117, 150, 153–4, 204; population projection 102–3; potential cropland 115–16; production 40, 76, 78, 81–3, 92–3, 95, 202; as region 29–30; trade 90–1; yields 121–4
South Korea 41, 55, 86, 89, 119, 128, 185–7
Southeast Anatolia Project 122, 179
Soviet Union 50, 70, 117, 131, 159, 164, 167, 193 *see also* former Soviet Union
soybean 93, 94, 143, 160
Sri Lanka 181
stability, socio-political 52–3, 56, 147, 177–8, 180, 185
stocks 14, 17, 69–73, 75, 84–5, 91, 98–9, 109, 165, 181, 188
storage 11, 21, 83, 95, 161, 165, 172
Sub-Saharan Africa: birth rate 32, 50; calorie availability 36–9, 42; cereal availability 46; cereal production, consumption and trade 40–1, 44; consumption 47, 108–11, 134; demand/supply 125–7, 133–4; demographic and socio-economic data 32; economic growth 105; famine 71, 73–5, 171, 177; fertilizers 119, 156; and food aid 45–6; food security 54–5; future prospects 166, 169, 170–7, 200; harvest variability 88–90, 142, 206–7; hunger 37, 201, 203; imports 205–6; irrigation 118, 154; population growth 10; population projection 101–2; potential cropland 115–17; production 40, 75–6, 78,

81–3, 93–6, 163; production/population growth 86, 97, 202; as region 29–31; satellite remote-sensing 162; supply 48–9, 56; trade 90–1; yields 98–9, 121–4, 159
subsidies 83–4, 91, 95, 98–9, 163–4, 167, 197–8
Sudan 29, 39, 55, 74, 85, 122, 178–80
Sukhatme-Margen hypothesis 35–7
sunflower oil 94
supply 34–9, 48–9, 52, 56, 115–24 *see also* demand and supply
Syria 67, 87, 122, 178, 180

Taiwan 65
Tajikistan 193
Tanzania 39
technology 8, 148, 155, 158, 171, 199
temperatures, global 136–8, 140, 142
Thailand 55, 85, 139, 185
today, population and food 24–57; cereal production, consumption, trade and aid 39–47; data sources and quality 24–8; definition and extent of hunger 33–9; dividing the world for global assessment 28–33; food security 51–5; population growth and food supplies 47–51
tomatoes 160
trade 40–1, 44, 89–91, 95–6, 126, 134, 141, 171, 178, 180, 185, 202, 204; liberalization 163–6, 167, 182, 200, 206
transition, demographic 48–50
transportation 11, 21, 95, 161–3, 172, 174
trends, population and food 1, 18, 58–99; changes in harvest variability 87–9; food crises and famines 68–75; food production, availability and variety 91–5; food trade and aid 89–91; green revolution 64–8; regional cereal production 75–86; world cereal production 58–63
triticale 158
Tunisia 178–80
Turkey 29, 55, 85, 122, 147, 178–80

Uganda 175–6
Ukraine 140, 193, 196, 198
undernutrition 34–9, 47, 51, 97, 180, 206
United Nations (UN) 24–5, 28, 101–5, 144, 184, 186
United States (US) 29, 41, 44–5, 69–70, 73, 81, 82, 88–9, 97–9, 117, 131, 140, 146, 149–51, 156, 167, 178, 187, 191, 195–9
urbanization 32–3, 95–6, 101–2, 105–6, 174

vegetables 92, 94, 199
Venezuela 44, 90
Viet Nam 55, 185–7, 189–90

warfare 50–2, 74–5, 85, 147, 171, 176, 185, 193
water 13–14, 20, 28, 118, 167–8, 179; for agriculture 149–55; constraints 127, 178, 188; renewable 19, 87
waterlogging 150–1, 179
wheat 15–16, 26, 41, 70, 85, 94–5, 139, 158, 178, 181, 187, 197–8; HYVs 7, 64–5, 172
wind erosion 145
World Bank 153, 173
world food crises 17, 69–74, 97
World Trade Organization (WTO) 163, 205

Yemen 55, 85, 178
yields: decrease 14, 146–7; in Europe/former Soviet Union 123, 193–4, 196; in Far East 79, 123, 188–90; fertilizers and 16, 119–20; global warming and 140–1; HYV 64–5, 68; increase 5, 11, 20, 60–3, 75, 81–3, 96–9,

123, 203–4, 207; inputs and 155–6, 159–61, 167–8; irrigation and 117; land degradation and 145–7; in Latin America 78, 123, 191–2; level 19; in Middle East 79, 123, 179; in North America/ Oceania 79, 123, 197, 199; production and 59–63; projections of 125–34; in South Asia 78, 123, 182–3; in sub-Saharan Africa 78, 123–4, 170; trends 122–4
Yugoslavia 50, 193

Zaïre 39, 174
Zambia 173, 176
Zimbabwe 31, 175–6